TEACHER GUIDE

7th –8th Grade

Includes Stu...
Workshee...

Science

Quizzes & Tests

Intro to Meteorology & Astronomy

MASTERBOOKS
—CURRICULUM—

First printing: April 2016
Second printing: March 2017

Master Books®, P.O. Box 726, Green Forest, AR 72638

Master Books® is a division of the New Leaf Publishing Group, Inc.

ISBN: 978-0-89051-964-6

Unless otherwise noted, Scripture quotations are from the New King James Version of the Bible.

Printed in the United States of America

Please visit our website for other great titles:
www.masterbooks.com

For information regarding author interviews,
please contact the publicity department at (870) 438-5288

"I'm loving this whole line so much. It's changed our homeschool for the better!

—Amy ★ ★ ★ ★ ★

"Your reputation as a publisher is stellar. It is a blessing knowing anything I purchase from you is going to be worth every penny!

—Cheri ★ ★ ★ ★ ★

"Last year we found Master Books and it has made a HUGE difference.

—Melanie ★ ★ ★ ★ ★

"We love Master Books and the way it's set up for easy planning!

—Melissa ★ ★ ★ ★ ★

"You have done a great job. MASTER BOOKS ROCKS!

—Stephanie ★ ★ ★ ★ ★

"Physically high-quality, Biblically faithful, and well-written."

—Danika ★ ★ ★ ★ ★

"Best books ever. Their illustrations are captivating and content amazing!

—Kathy ★ ★ ★ ★ ★

Affordable
Flexible
Faith Building

Master Books® Curriculum

Table of Contents

About Our Creationist Authors

The New Weather Book: **Michael Oard** earned his master's degree in atmospheric science in 1973 from the University of Washington. He was a meteorologist with the National Weather Service beginning in 1973 and lead forecaster in Great Falls, Montana from 1981 to 2001. He has written numerous books related to the Ice Age, geology and the Great Flood.

The New Astronomy Book: **Dr. Danny R. Faulkner** has a B.S. in Math, M.S. in Physics, M.A. and Ph.D. in Astronomy from Indiana University. He previously taught physics and astronomy at the University of South Carolina — Lancaster, and is now on staff at Answers in Genesis and the Creation Museum.

Features: The suggested weekly schedule enclosed has easy-to-manage lessons that guide the reading, worksheets, and all assessments. The pages of this guide are perforated and three-hole punched so materials are easy to tear out, hand out, grade, and store. Teachers are encouraged to adjust the schedule and materials needed in order to best work within their unique educational program.

Lesson Scheduling: Students are instructed to read the pages in their book and then complete the corresponding section provided by the teacher. Assessments that may include worksheets, activities, quizzes, and tests are given at regular intervals with space to record each grade. Space is provided on the weekly schedule for assignment dates, and flexibility in scheduling is encouraged. Teachers may adapt the scheduled days per each unique student situation. As the student completes each assignment, this can be marked with an "X" in the box.

🕐	Approximately 30 to 45 minutes per lesson, three days a week
🔑	Includes answer keys for worksheets, quizzes, and tests.
📝	Worksheets for each section
📄	Quizzes and tests are included to help reinforce learning and provide assessment opportunities.
🔁	Designed for grades 7 and 8 in a one-year course to earn ½ science credit

Course Description

This is the suggested course sequence that allows one core area of science to be studied per semester. You can change the sequence of the semesters per the needs or interests of your student; materials for each semester are independent of one another to allow flexibility. In the semester on meteorology, students will learn about God's design of this complex world and its weather patterns that affect our lives every day. The semester on astronomy extends God's design to the universe itself, and how all creation declares the glory and power of God. The universe is beautiful and breathtaking in its scale, and the earth and vast expanse of the universe is a struggle to study, understand, or even comprehend in terms of its purpose and size. Now take an incredible look at the mysteries and marvels of earth's weather and the far reaches of space.

Students completing this course will:

- Investigate how clouds form and how to identify the different types
- Review how to read a weather map, and what our responsibility is to the environment
- Learn how to survive in dangerous weather
- Identify what we know and are still trying to discover about planets, moons, and comets within our own solar system
- Evaluate up-to-date astronomical data and concepts
- Explore the dynamics of planets, stars, galaxies, and models for the cosmology of the universe
- Discover the best ways to observe the heavens.

Special Note: High school students who take the course are expected to do a majority of the activities. The activities can be modified based on student interests and creativity, but should reflect an understanding of the core concepts being learned.

Suggested Optional Science Lab

There are a variety of companies that offer science labs that complement our courses. These items are only suggestions, not requirements, and they are not included in the daily schedule. We have tried to find materials that are free of evolutionary teaching, but please review any materials you may purchase. The following items are available from www.HomeTrainingTools.com.

The New Weather Book

   KT-WESTUDY Weather Study Kit

The New Astronomy Book

   KT-SPACEXP Space Exploration Kit

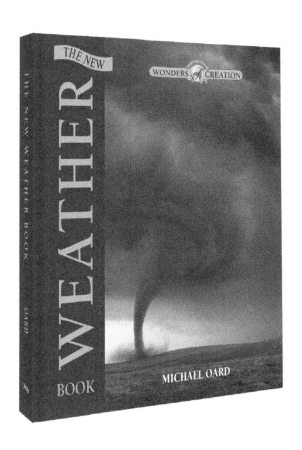

Meteorology Worksheets

for Use with

The New Weather Book

Words to Know (Definitions can be found in the Glossary in the back of this Teacher Guide.)

atmosphere

axis

carbon dioxide

climate

latitudes

nitrogen

oxygen

tide

Questions

1. How did Adam and Eve's first sin affect the weather?

2. Why can humans predict the weather?

3. Explain how weather affects your life.

4. Explain the anthropic principle in relation to the following terms:
   a. Tides:

   b. Seasons:

   c. Temperature:

   d. Atmosphere:

Words to Know

arid

barometer

condensation

dew point

Doppler radar

Questions

1. _____ is the momentary condition of the air.

2. Weather is composed of seven components. What are they?
   a.

   b.

   c.

   d.

   e.

   f.

   g.

3. Why is the sun called the weather engine?

4. _____ act like a blanket to keep the earth warmer at night.

5. What causes the earth's winds?

6. Describe what causes the Coriolis Force.

7. True or False: The Coriolis force moves in a clockwise rotation.

8. What causes the jet stream?

Activity:

Sketch a picture of the globe and jet streams on page 11 of *The New Weather Book*. Be sure to label the jet streams and the equator.

Words to Know

equator

low-pressure system

ice cap

meteorologist

precipitation

weather balloon

Questions

1. Where are storms usually found?

2. Give the memory tip to tell the difference between a cold and warm front.

3. Give two types of weather observations and how they are taken.
   a.

   b.

4. Draw and label the four types of weather fronts.

   a.

   b.

   c.

   d.

5. Why are weather forecasts sometimes incorrect?

6. Explain how general circulation causes climate.

7. What is the general circulation?

8. The earth has a total of _____ average circulations.

Activities (Note: Always ask a parent or teacher before using the Internet for research.)

Write a report on your weather. Describe the climate and features of the area that you live in and what affects your weather. (You may want to contact a local station and ask if you can interview a weather person.)

Write a report on the history of meteorology.

Write a report on NOAA. Describe what NOAA does, including the various components.

Produce a weather show detailing weather extremes and strange facts. (You may use information found on pages 18–19 of *The New Weather Book* but be sure to find some of your own.) Make it fun. Gather your friends and family to listen to your television show! Hint: Write notes on note cards as reminders of facts you want to give.

Words to Know

cirrus clouds

cold front

convection clouds

evaporation

fog

humid

Questions

1. Where does all the water for rain and snow come from?

2. Describe the water cycle.

3. True or False: The water cycle provides food and nutrients for sea life.

4. According to the salt levels in the ocean, what can we conclude about the age of the ocean?

5. Name the three types of clouds.

   a.

   b.

   c.

6. Which two cloud classifications are made of only water drops?

   a.

   b.

Activities (Note: Always ask a parent or teacher before using the Internet for research.)

Study and sketch the cloud types on page 25. Hang your drawing up to help remember the types and heights of clouds.

Record and sketch the types of clouds seen in your sky every day for a week. Describe what kind of weather accompanied or followed each type of cloud.

Write a report on Luke Howard's cloud categories

Words to Know

relative humidity

stratus clouds

thunderstorm

warm front

water vapor

Questions

1. What is a warm front?

2. Most clouds and precipitation are formed in areas of _____ air in the atmosphere.

3. What is the weather saying that sailors, farmers, and other people coined that was based on the repeating pattern they noticed of clouds and precipitation?

4. Describe the differences between warm fronts and cold fronts.

5. If a faster-moving cold front catches up to a warm front, it is called an _____ front.

6. What is the most dreaded type of cold front? Where is it located? Why is it so dreaded?

7. The _____ determines the most amount of vapor the air can hold.

8. What is the most common way for fog to form?

9. What is evaporation fog?

10. Describe how fog forms in mountainous or hilly terrains.

11. An _____ is where the temperature near the surface of the ground is colder than the air above it.

Activities

Sketch the two diagrams shown on pages 28–29. Describe what is happening in each.

Sketch the fog diagram shown on page 30.

Words to Know

downdraft

electricity

electrons

Questions

1. Use these numbers to answer the questions: 1.4, 20, 40 to 50, 242, 1,800
   a. The number of thunderstorms a year in Kampala, Uganda:

   b. The number of thunderstorms occurring around the earth at any one moment:

   c. The number of lightning bolt strikes across the earth every second:

   d. _____billion lightning bolts strike worldwide every year.

   e. The number of minutes it takes for a thunderstorm to produce a week's worth of electricity for a large city:

2. True or False: Stratus clouds form thunderstorms.

3. List the three conditions necessary for a thunderstorm:
   a.

   b.

   c.

4. What strengthens the updraft?

5. What causes the cumulus cloud to mushroom upward and transform into a huge towering cumulus or cumulonimbus cloud?

6. When does thunder and lightning usually start?

7. Does hail strengthen or weaken a thunderstorm?

8. Although thunderstorms can be dangerous, they are also a blessing. What are some of the benefits of thunderstorms?

    a.

    b.

    c.

    d.

    e.

Activities

Sketch The Anatomy of a Thunderstorm on page 35.

Write a report on thunderstorms.

Words to Know

static electricity

updraft

latent heat

Questions

1. Who was the first person to demonstrate that a thunderstorm generates electricity?

2. List the safety rules you should follow during a storm:

3. True or False: Scientists are certain about the origin of lightening.

4. The _____ charged bolts are the most dangerous.

5. Thunder is created when lightning heats surrounding air molecules to what temperature?

6. What does lightning sound like when it is near? What does it sound like when it is farther away?

7. How can you determine your distance from a thunderstorm?

Activities:

Write a report on lightning.

Create a Lightning Safety poster using the tips on page 37.

Sketch the three cloud images that explain the process of lightning on page 38.

Words to Know

flash flood

hailstones

supercell

tornado

Questions

1. Of the _____ thunderstorms that occur in the United States each year, _____ brings damaging winds, large hail, tornadoes, and flash floods.

2. True or False: Dangerous thunderstorms are examples of how Adam and Eve's rebellion affected nature, which is groaning for redemption.

3. Distinguish between a regular thunderstorm and a severe thunderstorm and list the conditions required for a severe thunderstorm.

4. _____ _____ occur when slow-moving thunderstorms drop an unusual amount of water on a small area and can trigger catastrophic _____ _____.

5. _____ and _____ are two damaging results of thunderstorms.

6. What determines the size of hail?

7. How large was the largest hail stone on the official world record? Where and when did it fall?

8. What do weathermen call the chaotic winds near the ground that are a danger for airplanes?

Activities

Write a report on a state where the most severe thunderstorms occur. Include information on why that state has more severe thunderstorms and statistics. Also find out what special precautions they take.

Create a Flash Flood Safety poster for your family.

Sketch the strong wind currents in the cloud shown on page 43.

Write a report on hail.

Words to Know

dust devil

supercooled drops

funnel cloud

Questions

1. At what point does a tornado officially become a tornado?

2. When a tornado touches down on water what is it called, where do they most frequently occur, and where are the largest number of them?

3. What factors are present when a supercell is formed?

4. Although scientists do not know the exact _____ that forms tornadoes, they now think they understand the _____ _____ of tornadoes.

5. Name more things scientists do not understand about tornadoes?

6. Predicting tornado _____ and _____ remains a challenge.

7. Storm chasers have learned that a tornado forms in a special spot under the thunderstorm where there is little _____ or _____, in the _____ part of the storm cloud.

8. All strong tornadoes form in a low overhanging cloud that lies just below the thunderstorm. This is called a _____ cloud.

9. What is the difference between a tornado watch and a tornado warning?

Activities

Sketch the differences in appearance of a funnel cloud, dust devil, and water spout.

Sketch the tornado illustration on page 46.

Write a report on the Joplin, Missouri, tornado that took place on May 21, 2011.

Words to Know

microburst

waterspout

tornado alley

Questions

1. Tornadoes are a regular feature for many who live in states within the _____ _____ region of the United States. They occur more frequently in the central parts of the U.S. between the _____ Mountains and the _____ Mountains.

2. The _____ scale classifies tornado strength from 0 to 5.

3. Describe the most dangerous tornadoes.

4. List the 4 tornado safety tips.

   a.

   b.

   c.

   d.

5. Tornadoes are _____, even in the midwestern United States.

6. What do meteorologists use to detect tornadoes and why are they effective?

Activities

Create a Tornado Safety Poster for your family based on the tips on page 48.

Write a report on tornadoes.

Words to Know

Intertropical Convergence Zone

tropical depression

monsoon

Questions

1. Name each geographic location's tropical weather phenomenon:

   a.

   b.

   c.

2. What the two processes that create the rainy areas in the tropics?

   a.

   b.

3. What causes most of the rainfall records on earth?

4. Name and describe the three type of storms in the tropics:

    a.

    b.

    c.

5. How do hurricane hunters work?

6. Researchers have learned that most hurricanes form after the ocean water warms up past _____°F.

7. The most violent band with the heaviest rain is the _____, which surrounds the eye or center of the hurricane.

8. When a hurricane moves inland or over colder water, it quickly _____ and _____ _____ .

Activities

Draw a sketch of the Anatomy of a Hurricane on page 53.

Write a paper on the monsoons.

Words to Know

storm surge

tropical storm

typhoon

Questions

1. Fill in the blanks using these numbers: One, two, three, four, five, six, seven, eight

   Generally, _____hurricanes a year form in the North Atlantic tropical zone. Only about _____ of these storms travel as far north as the east coast of the United States. The other _____hurricanes are pushed into the Caribbean Sea or into the Gulf of Mexico.

2. Since the 1920s, has the average number of deaths in the United States from hurricanes increased or decreased?

3. Since the 1920s, has property damage in the United States from hurricanes increased or decreased?

4. True or False: Since 1979, scientists have given hurricanes male and female names alternately and alphabetically, according to a list of names determined five years in advance.

5. Much _____ from old ships still lies on the ocean bottom, thanks to hurricanes.

6. True or False: Hurricanes, like tornadoes, are difficult to recognize and forecast with modern technology like satellites and Doppler radars.

7. What hurricane characteristic causes about 90 percent of hurricane-related deaths?

8. _____ and severe _____ usually happen after the hurricane moves onshore. These storms occur mostly around the hurricane's edge.

9. Identify the storm and the year:

    Andrew in Florida; Bangladesh; Bangladesh; Galveston, Texas; Iniki in Hawaii; Katrina in the Gulf; Sandy

    1900; November 12, 1970; 1991; 1992; August 24, 1992; 2005; August 2005; October 2012

    a. The most unusual hurricane to hit the United States; a "Superstorm":

    b. The deadliest natural disaster in U.S. history, killing 7,200 people:

    c. Developed in the Pacific Ocean off the Central American coast, causing $2 billion in damage:

    d. The costliest natural disaster in the United States at $108 billion and one of the five deadliest hurricanes with 1,833 fatalities.

    e. Record-breaking year for hurricanes, with 28 tropical and subtropical storms; 15 becoming hurricanes with five Category 4 and four Category 5:

    f. The worst hurricane in the 20th century that killed 300,000 to 500,000 people, leaving most of the remaining population homeless.

    g. Wrecked 60,000 homes, left 200,000 people homeless, damaged estimated at $25 billion, killed 23 in 2 states, caused 61 tornadoes and 177 severe storms, but early warning saved many lives.

    h. Killed around 138,000 people:

Activities:

Write a report on hurricane Katrina, Andrew, and/or Sandy.

Create a Hurricane Safety Poster for your family using the tips on page 59.

Words to Know

ice storm

sleet

wind chill factor

blizzard

Questions

1. What replenishes the rivers, streams, reservoirs, and groundwater in the springtime?

2. In a four-season climate, what causes cold and warm seasons?

3. Are thunderstorms more or less likely to occur during autumn?

4. In the Northern Hemisphere, when does the hurricane season end?

5. As temperatures drop, _____ and _____ are more frequent, usually lasting a day or two and bringing large amounts of moisture. _____ decides which it will be.

6. _____ usually falls in the northern section of North America, Europe, and Asia. _____ generally falls farther south, closer to the equator. The west coasts of the United States, Canada, and Europe rarely experience _____.

7. Why did God make snow white?

8. In what year did the Donner Party start their journey from Illinois?

9. Where were they headed?

10. As a result of bad decisions and slow going, they tried to cross the high _____ _____ _____ in late _____ .

11. What happened shortly after they left that forced them to camp near Truckee, California?

12. Of the _____ people who started in Illinois, only _____ survived.

Activities

Sketch the chart showing the earth's movement around the sun on page 61.

Write a report on the Donner Party.

Words to Know

avalanche

ice jam

Northeaster

Questions

1. Why are most winter rainstorms in the southern United States a blessing?

2. What sometimes happens to homes that are built on California hillsides after heavy rains?

3. Snowstorms are _____ and _____, but if people do not know their _____ they can be injured or die.

4. A blizzard describes blinding snow falling at what speed?

5. Give one example of how sin affects snow and winter storms.

6. When did the Storm of the Century occur and where did it strike?

7. A huge _____ developed in the northern Gulf of Mexico then curved northeast up the east coast.

8. How often does this type of storm hit?

9. When was the last storm of this size?

10. The 1993 storm was a powerful _____ — a storm that moves northeast along the east coast.

11. The Storm of the Century caused 15 _____ and severe _____ in Florida which _____ 44 people.

12. The wind and low pressure caused a 12-foot-high _____ _____ to hit the east coast.

13. When the snow stopped, it totaled _____ inches deep in Mount LeConte, TN.

14. The highest winds gusted to _____ mph at Grand Etang, Nova Scotia.

15. Describe in detail the damage from the Storm of the Century.

Activities

Write a report on the dangers present in California during the winter.

Sketch the area affected by the Storm of the Century shown on page 64.

Write a report on the Storm of the Century.

Words to Know

frostbite

hypothermia

inversion

Questions

1. Cloud temperatures ranging from 28–29°F causes _____, while cloud temperatures ranging from 32–34°F causes _____.

2. Winter storms can be _____ for a number of reasons. _____ and _____ can occur.

3. What are the most susceptible parts of the body to frostbite?

4. Describe hypothermia.

5. True or False: Someone showing signs of hypothermia should be wrapped in blankets and taken to the hospital immediately.

6. _____ percent of all winter deaths are the result of accidents involving people driving on icy and dangerous roads. Another _____ percent of storm-related deaths happen to people caught outside in storms with no shelter available.

7. _____ _____ are caused by a strong temperature difference between the tropics and the mid and high latitudes.

8. Who issues watches and warnings for winter storms, blizzards, and ice storms?

Activities

Sketch the three types of winter precipitation shown on page 65. Include the descriptions.

Write a report on hypothermia and frostbite.

Create a Winter Storm Safety Poster for you family based on the tips on page 67

Words to Know

ball lightning

chinook winds

Santa Ana winds

St. Elmo's fire

Questions

1. Mount Waialeale, on the northwestern island of Kauai, has one of the _____ yearly rainfalls in the world. It receives almost _____ _____of rain annually.

2. In Washington state, you can start from a _____ _____ , travel east over the Cascade Mountains, and enter into a _____ — all within several _____.

3. What part of the Olympic Mountains receive so much rain and snow the area is considered to be a rain forest?

4. The greatest snowfall in any one year in the world is _____ _____ at Paradise Ranger Station, located at about 5,000 feet elevation on _____ _____ in the Cascade Mountains.

5. By the time you reach Yakima or Pasco, Washington, it's practically a _____. Yakima receives only _____ _____of rainfall a year.

6. How do mountain ranges affect weather and climate?

Activities

Write a report on the affects mountains have on climate.

Words to Know

trade winds

foehn winds

rain forest

lake-effect snowstorm

Questions

1. How is foehn pronounced?

2. What are foehn winds?

3. Where do most foehn winds occur and when?

4. Near the mountains, a wall of _____-like clouds are observed. These clouds are called the
_____ _____ .

5. What part of the world refers to foehn winds as chinook winds?

6. Why do tens of thousands of elk and deer move from their mountain home to spend the winter on the eastern slopes of the Rocky Mountains and the high plains?

7. How can chinook winds be dangerous?

8. Describe Santa Ana winds, including location.

9. What causes St. Elmo's Fire?

10. A lake-effect winter storm hit _____, NY, in November _____ and brought up to _____ _____ of snow on the area.

Activities

Sketch the diagram of a foehn wind shown on page 70. Include the explanation.

Write a report on ball lightning.

Write a report on the Great Lakes winter weather.

Words to Know

bogs

environment

permafrost

thermometer

Questions

1. What kinds of fossils have scientists found in the Sahara Desert?

2. How do we know humans lived in the Sahara with these animals?

3. True or False: Scientists have found animal fossils in places like Siberia that prove that the climate must have been different in the prehistoric past.

4. Why is the study of the past not science?

5. Explain how a scientist might determine the climate of the past.

6. Contrast the creation-Genesis Flood model of the past with the evolution-uniformitarian model of the past.

7. How can scientists, looking at the same data, draw completely different conclusions?

Activities

Read the account of creation in Genesis. Then look up and read the account of Noah's Flood. Write a summary of both accounts.

Words to Know

Ice Age

Medieval Warm Period

uniformitarianism

model

Questions

1. Most rocks on the surface of the earth are _____ rocks.

2. What can explain how sedimentary rocks were laid down quickly?

3. Describe one explanation the creation-Genesis Flood model gives for the discovery of warm-climate fossils in cold climates like Siberia.

4. What tells us the pre-Flood climate was warm?

5. _____ is the compressed accumulation of trees and plants that have been transformed by higher temperatures.

6. Although estimates vary, it seems like there was about 10 times as much carbon in the form of coal as the current plants and trees on the whole earth. This means there was 10 times the amount of _____ on the earth before the Flood.

7. How do we know the Ice Age followed the Flood?

8. What two requirements were present after the Flood to create an Ice Age?

9. After the Flood, enormous amounts of _____ _____ rose from the warm ocean and blew onto the _____ continents.

10. The Ice Age only lasted about _____ years.

Activities

Sketch the map of the maximum extent of the Ice Age found on page 79.

Write a report on how the biblical Flood caused the Ice Age.

Words to Know

sedimentary rock

water vapor

Little Ice Age

Questions

1. Scientists who believe in the evolution-uniformitarian model have difficulty accounting for the Ice Age. More than _____ theories have been invented to try to account for the Ice Age.

2. Both the _____ _____ and the _____ _____ can explain the abundant evidence of lakes and rivers that once existed in areas that are now desert.

3. During the Ice Age, the much wetter climate would maintain the lakes with streams and rivers for at least several _____ years.

4. Toward the end of the Ice Age, the _____ would cool, and the climate of Siberia would become colder and _____.

5. By relying on _____ processes as the key to the past, the evolutionary-uniformitarian model is _____ to explain these climate mysteries.

6. When was the Medieval Warm period?

7. When was the Little Ice Age?

8. When was the Maunder minimum?

Activities

Write a report on the climate during the Ice Age, the Medieval Warm Period, and the Little Ice Age.

Words to Know

El Niño

fossil fuels

ozone

Questions

1. The _____ is the average weather at a location.

2. Climate can _____; it does not stay _____.

3. Large volcanic eruptions and fewer number of sunspots, can cause _____ temperatures.

4. Where is the El Niño current located and why is it named for "the boy" or "Christ child?"

5. How does El Niño affect the climate?

6. The Indian Ocean El Niño occurs _____ with the Pacific El Niño.

7. Two of the major cycles are the _____ _____ Oscillation and the _____ _____ Oscillation.

8. True or false: The cause of the Pacific Decadal Oscillation is very well known.

9. What causes a climate fluctuation in the North Atlantic?

10. True or false: Scientists are not sure whether the NAO is controlled by changes in ocean properties or if the atmosphere changes the ocean properties.

Activities

Sketch the El Niño charts and explanations found on page 84.

Write a report on the effects of El Niño.

Words to Know

plankton

pollution

greenhouse (effect) warming

Questions

1. True or False: Greenhouse gasses keep the earth warm, but an increase in carbon dioxide leads some scientists to think the earth will become much warmer in the future.

2. What Bible verse should guide Christians when considering the facts about global warming?

3. True or false: All scientists agree that global warming is a serious threat and we need to act now.

4. True or false: The polar vortex is a result of global warming and is an unusual occurrence. .

5. Name one of the errors that affects temperature records.

6. _____ show that deforestation estimates in places like Brazil have been exaggerated.

Activities

Sketch the radiation chart on page 86 and include the explanations.

Write a report on global warming.

Words to Know

ultraviolet light

climate

La Niña

heat-island effect

Questions

1. Despite claims that global warming has led to an increase in dangerous storms, scientists have found that there has been no change in what seven weather events?

2. Give at least one example of negative changes in climate caused by man.

3. It is important to study the data long and carefully before we make _____ in our world that would greatly affect our political and _____ system.

4. _____ is formed in the stratosphere when sunlight strikes oxygen.

5. How does the ozone layer benefit the earth?

6. The ozone hole has more _____ backing than the predictions of _____ _____.

7. What did the 1987 Montreal Protocol treaty ban to protect ozone and prevent growth of the ozone hole?

8. In trying to figure out the cause of the ozone hole, scientists have discovered that nature is very _____.

9. They found that the amount of _____ in the stratosphere goes in cycles.

10. The evolutionary-uniformitarian model advocates have a problem with their theory. They cannot evolve life _____ _____ in the atmosphere, and they cannot do it _____ _____.

Activities:

Sketch the globe and explanation of ozone on page 90.

Write a report on Ozone.

Words to Know

creation

environmentalist

Theory of Evolution

Questions

1. What Bible verse tells us that we are responsible for taking care of God's creation?

2. True or False: We should not be concerned about topics like climate change, the ozone hole, and air pollution.

3. We are _____ of God's creation.

4. True or False: Living simpler lives, as some scientists suggest, will prevent and solve environmental problems.

5. How should Christians approach environmental problems like pollution?

Activities

Write a report on how Christians can positively affect the environment.

Words to Know

Pantheist

steward

Questions

1. A _____ _____ is used to show the direction of the wind. It consists of four directional pieces that represent north, south, east, and west. One part remains _____, while the _____ — which moves freely — turns to indicate wind direction.

2. _____ is used in thermometers because it is a liquid at room temperature and produces a noticeable change when the temperature changes.

3. A _____ _____ gathers and measures the amount of precipitation over a specific period of time.

4. _____ are used to find out the atmospheric pressure.

5. What is atmospheric pressure?

6. If the atmospheric pressure rises quickly in a short time, this could correlate to _____ skies.

Activities

Use a weather vane, thermometer, rain gauge, and barometer to study the weather. Keep the information in a journal.

Astronomy Worksheets

for Use with

The New Astronomy Book

Special Course Activity Options

(NOTE: These can be done in combination with or rather than activities on each worksheet as noted on the daily schedule.) All project options will have the opportunity to score 100 points for their selected project. NASA, creation ministries, and libraries can be important sources of information for the projects of choice. Please be aware when searching other space-related sites that most do not include a biblical worldview.

Project notebook: This is a way to record results from the various suggested activities on *The New Astronomy Book* worksheets. NOTE: Grading should be based on the quality of research, coherence of presentation, good essay structure, and attribution of sources used.

Creative or Conceptual Projects: Artistic or engineering-focused; these are projects that include drawing or conceptualizing based on limited information. Could include various forms, for example:

- Drawing a series of images of imaginative colonies and facilities on the moon or Mars, ship designs that consider use of alternative renewable energy sources as fuel, exploration of potential commercial aspects of space-related activities.

- Student may choose to incorporate space-themed selections as part of a separate art course, or choose to imagine and draw structures, buildings, or dwellings related to space travel or colonization. (For example: imagine and draw a space library with no printed books, or greenhouses on Mars.)

- Younger students may want to fashion a children's story that they write and illustrate related to travel and study on another planet. Or they may do simple drawings.

Final term paper: You may assign a final term paper for the student over an aspect of the course materials. The subject can be of your choosing based on each student's interests or abilities, or you can do one of these suggested topics:

Historical

- History of man's race to space
- The story of NASA's formation and operations
- A landmark space-related achievement
- Development of telescopes or other technology aiding in man's study of stars

Cultural

- Which countries have achieved reaching space and why? Why have some not?
- What areas of possible dispute arise in terms of the space race? (i.e., military functions, claiming of resources, potential monopolies by one or a small number of countries)?

Scientific

- Write an 8- to 10-page biography of a scientist, astronomer, or astronaut mentioned in the book
- Discuss any theory noted in the book or related to space exploration or understanding the universe

Futuristic

- Imagine how a space-based governing body would function when the focus is no longer on individual countries or planets.
- What are the biggest challenges of manned exploration in distance space, and what solutions can you imagine to solve them?
- What viable reasons can you discuss that would justify spending vast amounts of money to colonize other areas of space or seek possible life in other places in space despite having no evidence that it exists, especially with the challenges already facing mankind on earth?

Words to Know

astronomical

comet

astrophysics

astronomy

spectroscopy

Questions

1. According to verses 14–15, what were the three purposes of stars?

2. What were the two great lights — the one that governs the day and the one that governs the night?

3. Verses 17–18 echo which of the other verses in the passage?

4. In verse 18, how does God describe this part of creation?

5. On what day of the creation week did these events occur?

Activity:

Take a look at the night sky!

Go out this evening just after dark. Mark your position on this drawing in relation to your house or a tree, and then draw in the moon and the constellations you recognize. Wait a few moments for your eyes to get used to the dark, and then note three other things that catch your attention in the night sky.

Hint! You can go to the following link: http://nightsky.jpl.nasa.gov/planner.cfm. You will find some helpful information and other links — including information on how clear it will be for night viewing and even a link for a site that allows you to download monthly evening sky maps for free. There are also free apps that help you identify what you see in the night sky.

You can also use the charts on pages 90 and 91 to help identify constellations.

Words to Know

constellations

axis

celestial

circumpolar

revolution

pagan

retrograde motion

Questions

1. Describe how stars move in the north near the North Pole.

2. How do circumpolar stars move in the Southern Hemisphere, and does the South Pole have a main star like Polaris is in the north?

3. Is Orion a winter or summer constellation in the Northern Hemisphere? Which would it be in the Southern Hemisphere?

4. What is a NEO? Can you give a recent example of a NEO that was mentioned on the news?

5. Is the study of the constellations a result of discoveries by modern science and space probes?

6. Which has the faster orbit cycle, the moon or the sun?

7. How long is the rotation of the sun? How long is the rotation of the moon?

8. Is rotation or revolution the circular motion around an axis that passes through the center of a body, such as a planet or moon?

9. List the five planets that appear as bright as stars in the sky.

10. The light of which two planets is too faint to be seen with the naked eye, and was not discovered until the invention of telescopes?

Activity:

Try to create a model out of things readily available in your home that will allow you to demonstrate retrograde motion. Remember, you can be as creative as you want to be, but it doesn't have to be a costly activity. For example, it could be done with balls, marbles, or even small rocks. Use chalk on a piece of cardboard to trace the path of objects and demonstrate retrograde motion.

Words to Know

unmanned

maria

highlands

lunar

impact basins

asteroids

bombardment

Questions

1. Compared to earth, how large is the moon?

2. What is the radius of the moon's orbit in miles?

3. Name one of three NASA space programs that focused on developing manned spaceflight.

4. What landmark of human achievements in space occurred on July 20, 1969?

5. What is synchronous orbit? Does the earth or the moon exhibit this orbit?

6. Who was the first person to see craters on the moon?

7. What two theories were debated in trying to explain the craters on the moon?

8. Why are some areas of the moon's surface darker than others?

9. How do evolutionists use EHB and LHB to explain the moon?

10. How does the moon produce light?

Activity:

Find a calendar that marks the phases of the moon. Highlight the following moon phases on this calendar, marking up six different months:

First Quarter
Gibbous
Waxing Gibbous
Full Moon
Waning Gibbous
Third Quarter
New Moon

Discover if the phases always land on the same days of the month. Is there a set number of days between the same phases in consecutive months? Is there a set number of days between each of the various phases within a given month?

Words to Know

Fill in the missing word in the paragraph below using the words from this list:

sun   high   full   lunar   spring tide

difference   neap tide   quarter   small

The _____ also produces tides, but because it is so much farther away than the moon, its tides aren't nearly as high. At new and _____ moon, both _____ and solar tides work together, and we say that this is _____ _____ . This name refers to how much the tides leap from very low levels at low tide to very high levels at high tide. On the other hand, at the _____ phases, the lunar and solar tides compete. This is a _____ _____ , and low tide is large, but the _____ between high and low tide is small at neap tide.

Questions

1. How does gravity from both the moon and the earth impact one another?

2. Is the moon's gravity pull on earth evenly distributed?

3. What is a lunar eclipse?

4. What is the only time a lunar eclipse can occur, and why?

5. What is the other name for the earth's shadow?

6. What clues helped lead ancient astronomers to realize that the earth was round?

7. Did the early church or Christians really teach that the world was flat?

8. During total solar eclipses, what are the corona and the prominences?

9. Why do the sun and the moon appear to be the same size during a solar eclipse?

10. What is the shape of the moon's orbit?

Activity:

In a short, one-page essay, explain how the sun and the moon and stars help to mark the passage of time.

Go to the library and see if you can find a book about astronauts Neil Armstrong, Buzz Aldrin, or Michael Collins. See if you can discover the training they had to undergo in order to become astronauts. Imagine space travel today with all of the amazing technological advances. What kinds of skills do you think you would need in order to become an astronaut today? Write down five possible skills you feel would be needed. Visit www.nasa.gov and search for astronaut training to see if any of your guesses are part of the program!

Words to Know

satellites

planet

minor planets

nucleus

ellipse

coma

Questions

1. Which of the planets of our solar system is the smallest one?

2. Why is Pluto no longer considered the smallest planet?

3. What are the names of the minor planets in our solar system?

4. List the planets of the inner solar system.

5. List the planets of the outer solar system.

6. What is the heart of a comet? What is it made of?

7. What is the name of the agency that controls the naming of objects in space?

8. What is the name of the comet that created a sensation by being viewed plowing into Jupiter in 1994?

9. What are three ways that comets are lost?

10. Why aren't the moons Ganymede and Titan considered planets even though they are bigger than Mercury?

Activity:

Go to NASA's webpage and search for the following comets:
    Hale-Bopp
    Halley

Why do you think these comets are so well-known? See what details you can find on each, and then pretend you are a reporter and write a short article on the comet of your choice.

Words to Know

Astronomical unit

Mass

Density

Rotation periods

Questions

1. Why was the astronomical unit created?

2. Using the chart on page 27, which planet has the longest counter clockwise (CCW) orbit in terms of days?

3. Which planet has the largest diameter?

4. Which two planets rotate backward or clockwise?

5. Which planet has the greatest density?

6. What do the orbits of the two planets that orbit clockwise tell us in terms of discounting the impact theory for explaining why they orbit directionally different than the other planets?

7. What are the two classifications of planets?

8. Which type of planet is closer to the sun?

9. Why type of planet has a lot of satellites and rings and is larger, with less density?

10. Why is Mercury so difficult to observe from earth?

Activity:

Study the poster included with *The New Astronomy Book*. Take a piece of paper, and see if you can re-create it by memory, drawing in planets and moons based on the appearance of the original poster. Notice the placement of the astronaut — this really shows just how little we have ventured out into our solar system over the last 50 years.

Words to Know

terrestrial

jovian

Galilean satellites

cryovolcanism

Questions

1. Which planet is sometimes called "earth's twin"?

2. Why did scientists assume there was water on Venus?

3. How does the presence of CO_2 in the atmosphere of Venus keep the planet hot?

4. Which planet has the largest known volcanoes in the solar system?

5. What do scientists think the planet Mars used to be like?

6. Which planet has a banded structure and the Great Red Spot?

7. What makes Io unique in our solar system?

8. Why do scientists feel Europa could harbor life of some kind?

9. How does the appearance of rings around Jovian planets point to a young universe?

10. Which of the planets rotate almost perpendicular to its revolution?

Activity:

Try to imagine the capabilities of a space probe designed to visit both Venus and Jupiter and their surfaces. What are some of the problems you think would have to be overcome in design and equipment to have a successful mission to gather data on both of these planets?

Words to Know

fission

fusion

potential energy

paradox

sunspots

photosphere

umbra

penumbra

magnetic polarity

Questions

1. What powers the sun?

2. What is the opposite of fusion?

3. Why can't gravitational potential energy be what is powering the sun in an evolutionary-based model of the universe forming?

4. Do scientists think earth was a lot colder in the past?

5. If nuclear fusion is powering the sun over millions of years, what would the earth be like today?

6. What is the "young faint sun paradox"?

7. From which part of the sun do we receive almost all of our radiation?

8. How can sunspots be used to tell how long it takes the sun to rotate?

9. How long is a sunspot cycle?

10. Why do sunspots appear in pairs?

Activity:

In the book you learned the following:

> During sunspot maximum, there is much magnetic activity on the sun. This magnetic activity can produce flares on the sun. A solar flare releases much energy into space, but most of the radiation is invisible to the eye. A solar flare also releases many charged particles. These charged particles amount to a gust in the solar wind. These charged particles are moving very fast, and they take a day or two to reach the earth.

> The charged particles interact with the earth's magnetic field to cause magnetic storms. Magnetic storms can interrupt electronic communications. They also can cause an aurora as the charged particles strike atoms high in the earth's atmosphere. Aurorae more often are called northern lights, at least in the Northern Hemisphere, because you generally see them in the northern part of the sky. In the Southern Hemisphere, people call aurorae the southern lights.

Go to the library or a trusted website and see if you can find various pictures of the Northern or Southern Lights. Then see if you can create an illustration that shows how they form, based on an excerpt from your book.

Words to Know

spectrograph

wavelengths

electromagnetic radiation

infrared radiation

microwaves

radio waves

spectroscopy

spectral lines

Doppler Effect

Questions

1. What are the two basic types of telescopes?

2. How is the size of a telescope determined?

3. Why do telescopes have to be so large to look into distant space?

4. Why do astronomers attach special cameras to their telescopes?

5. Do stars always have a consistent brightness?

6. What does ROYGBIV stand for?

7. Which light has wavelengths too short for us to see?

8. How does the atmosphere protect us on earth from electromagnetic radiation?

9. Why are scientists able to study radio waves from space from earth?

10. How does using the spectra of astronomical bodies help tell us what substances form these bodies?

Activity:

Write a one-page essay highlighting some pros and cons of funding things like the James Webb Space Telescope rather than using the money to solve problems here on earth.

Words to Know

parallax

Geocentric Theory

Heliocentric Theory

epicycle

retrograde motion

laws of planetary motion

physics

Questions

1. In what three ways did ancient Greeks discover the earth was round?

2. Who measured the earth's size in 200 B.C., and how was it measured?

3. Ancient Greeks could see that the sun moved once around through the stars in a year, but what two important questions did this knowledge create?

4. Why did many ancient people believe the geocentric theory?

5. Why did retrograde motion puzzle ancient astronomers?

6. How did Ptolemy help find a solution for the problem of retrograde motion in the geocentric model?

7. Which astronomer's book helped to convince both Galileo and Kepler that the heliocentric theory was correct?

8. What three discoveries by Galileo proved the geocentric theory was incorrect?

9. Who discovered three laws of planetary motion by studying the work of Tycho Brahe?

10. Who was the president that launched the race to the moon in 1961?

Activity:

Review the chart at the bottom of page 47 in your book. This lists a chronology of missions by U.S. Astronauts from 1961 to 1972. Choose one of the missions and visit www.nasa.gov and search for details on that specific mission to see if you can find historical data or images.

Words to Know

light years

absolute magnitude

parsec

binary systems

white dwarf stars

neutron star

black hole

supernovae

pulsar

Questions

1. Which type of mathematics is used for helping find the distance of stars from earth?

2. What is the name for the system created to measure the brightness of stars?

3. What is the brightness of the following:
   a. Sun: _____
   b. Full moon: _____
   c. North Star: _____

4. What is apparent magnitude?

5. What does the color of a star tell us?

6. What is the most common element in the universe?

7. What can scientists learn from studying binary stars?

8. Why are giant stars said to be very "thin"?

9. How do we know that black holes really exist when we cannot see them?

10. Do novae keep their new brightness?

Activity:

Go to www.nasa.gov and look for "Apollo 8: Christmas at the Moon." Then listen to this recording of the famous Christmas message from the crew of the Apollo 8 mission in 1968. They took turns reading from the Book of Genesis. The NASA link is: http://www.nasa.gov/topics/history/features/apollo_8.html#.U_UVcfldV8E.

Words to Know

extrasolar

exoplanet

main sequence star

transit

habitable zone

Questions

1. Are binary stars uncommon?

2. Why were extrasolar planets so hard to find?

3. How are extrasolar planets named?

4. Since the extrasolar planets are so difficult to see, what is an easier way to find them?

5. The more mass a star has, does it move more or less?

6. Do planets or stars have more mass?

7. How can transit help in the discovery of extrasolar planets?

8. What spacecraft has been used to find a very large number of extrasolar planets?

9. Why are scientists so interested in extrasolar planets?

10. How are extrasolar planets a problem for those who believe in evolution?

Activity:

Think about the different types of energy we use here on earth — electrical, nuclear, atomic, solar, wind, and others. Research different forms at the library in a book about energy if possible. Choose one form of energy and roughly draw out a plan for a space vessel powered by it.

Words to Know

star clusters

open clusters

globular clusters

interstellar medium

silicate

nebulae

dark nebula

reflection nebula

molecular clouds

Questions

1. What are field stars?

2. Do the stars in a cluster orbit around a common center of mass?

3. How many times is Pleiades mentioned in the Old Testament?

4. What are the two types of star clusters?

5. Why type of star cluster can have between 50,000 and a million stars?

6. Which type of star cluster has an irregular appearance?

7. What are four properties of the interstellar medium?

8. What is the Latin word for "cloud"?

9. What happens to the hydrogen gas in the process of ionization if a large cloud of hydrogen gas is around a hot, bright star?

10. Have astronomers ever witnessed the birth of a star?

Activity:

Choose an object you noted in the sky on Chapter 1, Worksheet 1. Now see if you can find that object in the night sky again. Once again, note the position of the object on the drawing below. Compare positions of the object since you first marked its location on the previous worksheet and where it is now.

Words to Know

spiral tracers

nebulae

spiral nebulae

Cepheid variables

spiral galaxies

irregular galaxies

Local Group

dwarf galaxies

Questions

1. What is the Greek word for "milk"?

2. Which astronomer wrongly concluded that the sun is at the center of the galaxy?

3. How big is the diameter of the Milky Way?

4. What kind of galaxy is the Milky Way?

5. Why are the spiral arms of galaxies a problem for people who think the universe is billions of years old?

6. What is the area surrounding the Milky Way galaxy's disk?

7. What invention helped astronomers determine that some nebulae were actually star clusters?

8. Up until 1924, what was the dominant theory on spiral nebulae?

9. Do galaxies evolve from one type of galaxy to another type?

10. What is the nearest galaxy cluster to earth?

Activity:

Try to create a model of the Milky Way galaxy using either modeling clay or a homemade version of air-hardening modeling clay (you can find a recipe at http://www.homeschooling-ideas.com/how-to-make-clay.html#airhardening). Use the images on pages 64–65 as a guide — you can even paint it if you wish.

Words to Know

general relativity

white hole theory

dimension

ASC solution

dasha solution

Ex Nihilo

Questions

1. What is the light-time-travel problem?

2. What are a couple of solutions to the light-time-travel problem?

3. List the names of at least three creation scientists working on theories to explain the light-time-travel question.

4. Why is the speed of light an issue in the ASC solution?

5. Use the need for plants to be eaten by Adam and Eve to explain the dasha solution.

6. If you could choose the first three places in the universe mankind would visit, what would they be?

List the three issues we must overcome to allow humans to visit distant places in the universe.

7.

8.

9.

10. Is the lack of a definitive answer for the light-time-travel issue one that should worry someone who believes in the creationist account of creation?

Activity:

Take four objects (balls, small rocks, etc.) and place them on the table. Now place a smaller object about two feet away from the others. On a series of index cards, make the following notes:

    Life expectancy: 30 years

    Life expectancy: 80 years

    Life expectancy: 100 years

Mark these with a green dot on the back.

On three other cards, make the following notes:

    Supplies: included, one-year supply only

    Supplies: created on board, basic supply elements

    Supplies: developed from substances as found in space

Mark these with a red dot on the back.

On three other cards, make the following notes:

    Powered: short-term solar energy from the sun

    Powered: new tech that creates dangerous waste

    Powered: clean, sustainable source

Mark these with a yellow dot on the back.

Turn all cards over and shuffle together. Randomly pick a card from each color and then turn them face up. See what the pros and cons of each selected set of cards will create. Be sure to do more than once. These are the current challenges of manned space exploration.

Words to Know

Hubble relation

redshifted

blueshifted

extra-galactic astronomy

standard candle

Questions

1. Which astronomer discovered the universe is expanding, and when was his discovery made?

2. What can we use Doppler shift to tell us?

3. What was Vesto Slipher's discovery?

4. If a galaxy has a greater redshift than another galaxy, this means what?

5. Do scientists fully understand the concept of the universe expanding?

6. Does the universe expanding mean that the big-bang theory is true, and creationism is false?

7. What is one of the biggest challenges to exploring space?

8. Are all potential space vehicles being designed and built by the government today?

9. Do warp-drive engines to power space vehicles currently exist like you see in popular movies?

10. The multiverse theory is a popular one today, but is there any evidence for this theory?

Activity:

Choose an astronomer you have read about and see what you can find out about his or her work and what generated his or her interest in studying astronomy. Biographies can often be found at the library or parent-approved websites. Many books are available for free download from Google books and Project Gutenberg.

Words to Know

radio stars

quasi-stellar objects

quasars

synchrotron radiation

synchrotron spectra

Questions

1. When was radio astronomy developed?

2. What color were the stars that produced so many radio emissions they were called "radio stars"?

3. What does QSO stand for?

4. Quasar 3C 273 was brighter than how many stars?

5. Do the spectra of quasars resemble stellar spectra?

6. What was the theory developed to try to explain how small quasars could be so bright?

7. Does synchrotron radiation provide evidence for or against the theory of supermassive black holes?

8. How much larger is the mass at the core of the Milky Way than that of the sun?

9. If most large galaxies have supermassive black holes at their core, then what does it mean when it is starved for matter?

10. When looking at the active galactic nuclei, what determines what we see?

Activity:

Consider making your own planisphere. You can download information at http://www.vcas.org/star-wheels.html. Study the charts on pages 90 and 91 and see if you can identify by memory at least six of the constellations in this drawing:

Words to Know

cosmology

filaments

voids

steady-state theory

continuous-creation theory

big-bang theory

cosmic microwave background

horizon problem

flatness problem

Questions

1. Are the heliocentric and the geocentric theories considered to be cosmologies?

2. What is a cosmologist?

3. How did the big-bang theory become the dominant cosmology among astronomers today?

4. What are the problems with Christians trying to think of the big bang as God's model for the creation of the universe?

5. What are two problems with the big-bang theory?

6. When expected temperature fluctuations in the cosmic microwave background did not match the ones theorized in the big-bang model, what happened?

7. How has the big-bang theory changed over time?

8. Why will it be a while before the big-bang cosmology model is abandoned?

9. The Johnson Observatory is located at what facility near Cincinnati?

10. What was the name of the NASA project for which Lyle T. Johnson developed a new class of telescopes?

Activity:

Go back and look at the pictures of stars and galaxies in this book. Imagine you are explaining the beauty and diversity of planets and stars within the universe to someone who has never seen pictures of them. Write three paragraphs to try and describe its vast, inspiring wonders, using descriptions and examples from the book.

Quizzes & Tests Section

Define: (5 Points Each Answer)

1. climate: _____

2. nitrogen: _____

3. oxygen: _____

4. precipitation: _____

5. cirrus clouds: _____

6. cold front: _____

7. evaporation: _____

8. stratus clouds: _____

9. warm front: _____

10. water vapor: _____

Short Answer Questions: (5 Points Each Question)

1. How did Adam and Eve's first sin affect the weather?

2. Why can humans predict the weather?

3. Why is the sun called the weather engine?

4. Draw and label the four types of weather fronts.

5. Why are weather forecasts sometimes incorrect?

6. Explain how general circulation causes climate.

7. Describe the water cycle.

8. Which two cloud classifications are made of only water drops?

9. Describe the differences between warm fronts and cold fronts.

10. What is the most common way for fog to form?

Define: (5 Points Each Answer)

1. electrons: _____

2. static electricity: _____

3. latent heat: _____

4. supercell: _____

5. tornado: _____

6. supercooled drops: _____

7. funnel cloud: _____

8. microburst:_____

9. monsoon: _____

10. tropical storm: _____

Short Answer Questions: (5 Points Each Question)

1. Thunder is created when lightning heats surrounding air molecules to what temperature?

2. How can you determine your distance from a thunderstorm?

3. Distinguish between a regular thunderstorm and a severe thunderstorm and list the conditions required for a severe thunderstorm.

4. What determines the size of hail?

5. At what point does a tornado officially become a tornado?

6. What is the difference between a tornado watch and a tornado warning?

7. The _____ scale classifies tornado strength from 0 to 5.

8. What do meteorologists use to detect tornadoes and why are they effective?

9. How do hurricane hunters work?

10. What hurricane characteristic causes about 90 percent of hurricane-related deaths?

Define: (5 Points Each Answer)

1. ice storm: _____

2. blizzard: _____

3. hypothermia: _____

4. inversion : _____

5. ball lightning: _____

6. St. Elmo's fire: _____

7. trade winds: _____

8. foehn winds: _____

9. environment: _____

10. thermometer: _____

Short Answer Questions: (5 Points Each Question)

1. Why did God make snow white?

2. Give one example of how sin affects snow and winter storms.

3. How do mountain ranges affect weather and climate?

4. What part of the world refers to foehn winds as chinook winds?

5. What is St. Elmo's Fire?

6. How do the Great Lakes affect the surrounding weather?

7. Why is the study of the past not science?

8. Explain how a scientist might determine the climate of the past.

9. Describe one explanation the creation-Genesis Flood model gives for the discovery of warm-climate fossils in cold climates like Siberia.

10. What two requirements were present after the Flood to create an Ice Age?

Define: (5 Points Each Answer)

1. El Niño: _____

2. ozone: _____

3. plankton: _____

4. pollution: _____

5. greenhouse (effect) warming: _____

6. ultraviolet light: _____

7. La Niña: _____

8. creation: _____

9. environmentalist: _____

10. steward: _____

Short Answer Questions: (5 Points Each Question)

1. How does El Nino affect the climate?

2. What Bible verse should guide Christians when considering the facts about global warming?

3. Name one of the errors that affects temperature records.

4. _____ show that deforestation estimates in places like Brazil have been exaggerated.

5. Despite claims that global warming has led to an increase in dangerous storms, scientists have found that there has been no change in what seven weather events?

6. What did the 1987 Montreal Protocol treaty ban to protect ozone and prevent growth of the ozone hole?

7. What Bible verse tells us that we are responsible for taking care of God's creation?

8. We are _____ of God's creation.

9. True or False: Living simpler lives, as some scientists suggest, will prevent and solve environmental problems.

10. How should Christians approach environmental problems like pollution?

| T | The New Weather Book
Concepts & Comprehension | Test 1 | Scope:
Chapters 1-11 | Total score:
____of 100 | Name |

Define: (2 Points Each Answer)

1. nitrogen: _____

2. oxygen: _____

3. cirrus clouds: _____

4. evaporation: _____

5. stratus clouds: _____

6. static electricity: _____

7. supercell: _____

8. tornado: _____

9. supercooled drops: _____

10. tropical storm: _____

11. ice storm: _____

12. blizzard: _____

13. hypothermia: _____

14. St. Elmo's fire: _____

15. trade winds: _____

16. foehn winds: _____

17. environment: _____

18. thermometer: _____

19. El Niño: _____

20. ozone: _____

21. greenhouse (effect) warmings: _____

22. ultraviolet light: _____

23. La Niña: _____

24. creation: _____

25. steward: _____

Short Answer Questions: (2 Points Each)

1. How did Adam and Eve's first sin affect the weather?

2. Why is the sun called the weather engine?

3. Explain how general circulation causes climate.

4. Describe the water cycle.

5. How can you determine your distance from a thunderstorm?

6. What is the difference between a tornado watch and a tornado warning?

7. What do meteorologists use to detect tornadoes and why are they effective?

8. How do hurricane hunters work?

9. Why did God make snow white?

10. Give one example of how sin affects snow and winter storms.

11. How do mountain ranges affect weather and climate?

12. What part of the world refers to foehn winds as chinook winds?

13. What is St. Elmo's Fire?

14. Why is the study of the past not science?

15. Explain how a scientist might determine the climate of the past.

16. What two requirements were present after the Flood to create an Ice Age?

17. What Bible verse should guide Christians when considering the facts about global warming?

18. Name one of the errors that affects temperature records.

19. Despite claims that global warming has led to an increase in dangerous storms, scientists have found that there has been no change in what seven weather events?

20. What Bible verse tells us that we are responsible for taking care of God's creation?

Applied Learning Activity: (5 Points)

21. Write the following words in their proper elevation on the diagram:

Nimbostratus, Stratus, Cumulonimbus, Cumulus, Altocumulus, Altostratus, Cirrostratus, Cirrocumulus, Cirrus, Cumulonimbus, Troposphere, Stratosphere

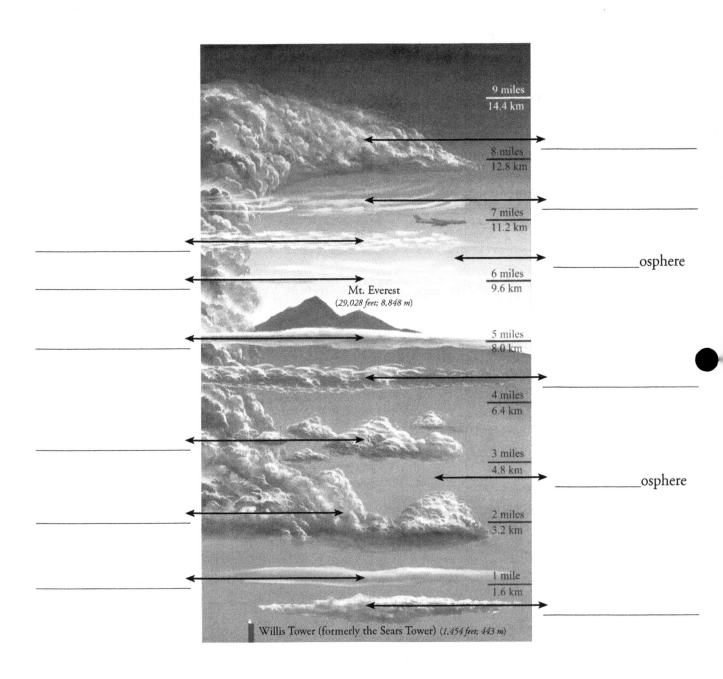

Words: (4 Points Each Answer)

1. astronomy: _____

2. spectroscopy: _____

3. comet: _____

4. axis: _____

5. retrograde motion: _____

6. maria: _____

7. highlands: _____

8. lunar: _____

9. satellites: _____

10. minor planets: _____

Questions: (4 Points Each Answer)

1. How do circumpolar stars move in the Southern Hemisphere, and does the South Pole have a main star like Polaris is in the north?

2. What is a NEO?

3. Which has the faster orbit cycle, the moon or the sun?

4. List the five planets that appear as bright as stars in the sky.

5. Is rotation or revolution the circular motion around an axis that passes through the center of a body, such as a planet or moon?

6. What is synchronous orbit? Does the earth or the moon exhibit this orbit?

7. Who was the first person to see craters on the moon?

8. How does the moon produce light?

9. How does gravity from both the moon and the earth impact one another?

10. What clues helped lead ancient astronomers to realize that the earth was round?

11. During total solar eclipses, what are the corona and the prominences?

12. What are the names of the minor planets in our solar system?

13. List the planets of the inner solar system.

14. List the planets of the outer solar system.

15. What are three ways that comets are lost?

Words: (4 Points Each Answer)

1. Astronomical unit: _____

2. Density: _____

3. Rotation periods: _____

4. Galilean satellites: _____

5. parallax: _____

6. fission: _____

7. fusion: _____

8. wavelengths: _____

9. electromagnetic radiation: _____

10. infrared radiation: _____

Questions: (4 Points Each Answer)

1. Why was the astronomical unit created?

2. Which two planets rotate backward or clockwise?

3. What are the two classifications of planets?

4. Which type of planet is closer to the sun?

5. Why type of planet has a lot of satellites and rings and are larger, with less density?

6. How does the appearance of rings around Jovian planets point to a young universe?

7. How does the presence of CO_2 in the atmosphere of Venus keep the planet hot?

8. Which planet has the largest known volcanoes in the solar system?

9. Do scientists think earth was a lot colder in the past?

10. From which part of the sun do we receive almost all of our radiation?

11. Why do sunspots appear in pairs?

12. What are the two basic types of telescopes?

13. How does using the spectra of astronomical bodies help tell us what substances form these bodies?

14. What three discoveries by Galileo proved the geocentric theory was incorrect?

15. Who measured the earth's size in 200 B.C., and how was it measured?

Words: (4 Points Each Answer)

1. light years: _____

2. absolute magnitude: _____

3. parsec: _____

4. white dwarf stars: _____

5. extrasolar: _____

6. transit: _____

7. habitable zone: _____

8. star clusters: _____

9. interstellar medium: _____

10. Local Group: _____

Questions: (4 Points Each Blank)

1. What is apparent magnitude?

2. What does the color of a star tell us?

3. What is the most common element in the universe?

4. What is the name for the system created to measure the brightness of stars?

5. Since the extrasolar planets are so difficult to see, what is an easier way to find them?

6. How can transit help in the discovery of extrasolar planets?

7. The more mass a star has, does it move more or less?

8. What are the two types of star clusters?

9. Why type of star cluster can have between 50,000 and a million stars?

10. Which type of star cluster has an irregular appearance?

11. Do the stars in a cluster orbit around a common center of mass?

12. How big is the diameter of the Milky Way?

13. What kind of galaxy is the Milky Way?

14. Up until 1924, what was the dominant theory on spiral nebulae?

15. What is the nearest galaxy cluster to earth?

Words: (4 Points Each Answer)

1. dimension: _____

2. *Ex Nihilo*: _____

3. general relativity: _____

4. Hubble relation: _____

5. blueshifted: _____

6. standard candle: _____

7. quasars: _____

8. synchrotron radiation: _____

9. cosmology: _____

10. filaments: _____

Questions: (4 Points Each Answer)

1. What is the light-time-travel problem?

2. List the three issues we must overcome to allow humans to visit distant places in the universe.

3. Which astronomer discovered the universe is expanding and when was his discovery made?

4. What can we use Doppler shift to tell us?

5. The multiverse theory is a popular one today, but is there any evidence for this theory?

6. What was the theory developed to try to explain how small quasars could be so bright?

7. When was radio astronomy developed?

8. If most large galaxies have supermassive black holes at their core, then what does it mean when it is starved for matter?

9. Are the heliocentric and the geocentric theories considered to be cosmologies?

10. What is a cosmologist?

11. What are two problems with the big-bang theory?

12. When expected temperature fluctuations in the cosmic microwave background did not match the ones theorized in the big-bang model, what happened?

13. What was the name of the NASA project for which Lyle T. Johnson developed a new class of telescopes?

14. How has the big-bang theory changed over time?

15. What was Vesto Slipher's discovery?

Define: (2 Points Each Answer)

1. Parsec: _____

2. Density: _____

3. Minor planets: _____

4. White dwarf stars: _____

5. Infrared radiation: _____

6. Absolute magnitude: _____

7. Blueshifted: _____

8. Electromagnetic radiation: _____

9. Astronomical unit: _____

10. Retrograde motion: _____

11. Axis: _____

12. Wavelengths: _____

13. Maria: _____

14. Parallax: _____

15. Highlands: _____

16. Fusion: _____

17. Light years: _____

18. Transit: _____

19. Star clusters: _____

20. Standard candle: _____

Questions: (2 Points Each Answer)

1. How do circumpolar stars move in the Southern Hemisphere, and does the South Pole have a main star like Polaris is in the north?

2. List the five planets that appear as bright as stars in the sky.

3. What is synchronous orbit? Does the earth or the moon exhibit this orbit?

4. How does gravity from both the moon and the earth impact one another?

5. During total solar eclipses, what are the corona and the prominences?

6. What are the names of the minor planets in our solar system?

7. List the planets of the outer solar system.

8. What are three ways that comets are lost?

9. Why was the astronomical unit created?

10. Which two planets rotate backward or clockwise?

11. What are the two classifications of planets?

12. How does the appearance of rings around Jovian planets point to a young universe?

13. How does the presence of CO_2 in the atmosphere of Venus keep the planet hot?

14. Which planet has the largest known volcanoes in the solar system?

15. Do scientists think earth was a lot colder in the past?

16. Why do sunspots appear in pairs?

17. How does using the spectra of astronomical bodies help tell us what substances form these bodies?

18. What is apparent magnitude?

19. What does the color of a star tell us?

20. What is the most common element in the universe?

21. What is the name for the system created to measure the brightness of stars?

22. How can transit help in the discovery of extrasolar planets?

23. The more mass a star has, does it move more or less?

24. Which type of star cluster has an irregular appearance?

25. Do the stars in a cluster orbit around a common center of mass?

26. What kind of galaxy is the Milky Way?

27. What is the light-time-travel problem?

28. What can we use Doppler shift to tell us?

29. If most large galaxies have supermassive black holes at their core, then what does it mean when it is starved for matter?

30. What are two problems with the big-bang theory?

Answer Keys

Weather Chapter 1: God Created

atmosphere – the body of gasses that surround the earth.

axis — an imaginary straight line through the center of the earth on which it rotates

carbon dioxide — a colorless, odorless gas formed during respiration, combustion, and organic decomposition

climate – the weather conditions that are particular to a certain area, such as wind, precipitation, and temperature.

latitudes — the distance north or south of the equator measured with imaginary lines on a map or globe

nitrogen — a naturally occurring element that is responsible for around four-fifths of the earth's atmosphere

oxygen — a colorless, odorless gas that is 21 percent of our atmosphere essential for plant and animal respiration

tide — a raising and lowering of the water in the oceans and seas caused by the gravitational pull of the moon. The sun causes some, but to a lesser degree.

1. When Adam and Eve disobeyed God's commands, they allowed evil to enter the world. Bad weather is a result of the presence of sin in the world.

2. God created the world with a perfect design and order, which allows us to predict hours of daylight, seasons, and weather.

3. Answers will vary.

4. a. Tides: God placed the moon at exactly the right distance to maintain oceanic tides, which prevent flooding and pollution.

   b. Seasons: God tilted the earth's axis; the earth's tilt and rotation creates seasons.

   c. Temperature: God created just the right amount of gasses in the air to make sure the earth isn't too hot or too cold.

   d. Atmosphere: God created an atmosphere around the earth that shields it from harmful rays and meteors

Weather Chapter 2: – What Causes Weather Worksheet 1

arid — a dry climate lacking moisture

barometer — a weather instrument used to measure the pressure of the atmosphere

condensation — the act of water vapor changing from a gas to a liquid

dew point — the temperature at which air becomes saturated and dew forms

Doppler radar — a special type of radar used to track severe weather by detecting wind speed and direction

1. weather

2. a. temperature; b. precipitation; c. wind direction and speed; d. visibility; e. water vapor in the air; f. cloud conditions; g. air quality

3. The sun is responsible for differences in heating around the earth. At night, infrared radiation cools the earth, and the sunrise warms the earth during the day.

4. Clouds

5. The difference between daytime sunshine and nighttime infrared cooling causes temperature differences between the tropics and polar latitudes. These temperature differences cause air pressure changes, which push the earth's winds.

6. Because of the earth spinning on its axis, air flow in the atmosphere is more complicated. The spin causes air to move to the right in the Northern Hemisphere and to the left in the Southern Hemisphere. This deflecting force on the air is called the Coriolis force.

7. False; the Coriolis force moves in a counterclockwise rotation

8. The jet stream is caused by the difference in temperature between the tropical and polar latitudes.

Weather Chapter 2: – What Causes Weather Worksheet 2

equator — an imaginary line dividing the Northern and Southern Hemispheres

low-pressure system — Warm, moist air that usually brings storms with strong winds. The air spirals counter-clockwise around a low center in the Northern Hemisphere and clockwise in the Southern Hemisphere. Because the air is spiraling toward the center of the low, it is forced upward, forming clouds and precipitation.

ice cap — an extensive covering of ice and snow

meteorologist — a person who interprets scientific data and forecasts the weather

precipitation — falling moisture in the form of rain, sleet, snow, hail, or drizzle

weather balloon — a balloon used to carry weather instruments into the atmosphere to gather data

1. The storm is usually found below the southwest wind of the jet stream.

2. Think of the triangles as icicles and the semicircles as blisters.

3. Surface Observation — Each weather station takes measurements of temperature, dew point, clouds, precipitation, pressure, and wind speed and direction.

   Upper Air Observation — This is done by weather balloons and taken twice a day, at noon and midnight Greenwich time in England.

4. Cold front: line with triangles. Warm front: line with semicircles. Stationary front: line with triangles on top and semicircles on the bottom. Occluded front: line with alternating triangles and semicircles on top

5. Meteorologists can only interpret weather maps created by computers, which are not perfect. Despite all the tools they have, meteorologists do not have a complete understanding of the earth and its atmosphere.

6. The general circulation is the average flow of air in various locations. The earth has six average circulations, which determines the climate in different areas.

7. The differences in heating across the earth and the weather cause an average flow of air called the general circulation.

8. six

Weather Chapter 3: – Water in the Atmosphere Worksheet 1

cirrus clouds — a high altitude cloud made of ice crystals that appears thin, white, and feathery

cold front — a boundary of cold air, usually moving from the north or west, which is displacing the warm air

convection clouds — clouds that occur in a rising up-draft, usually when the sun's radiation warms the earth. This causes the water vapor to condense.

evaporation — to change into a vapor such as the evaporation of water by the warming of the sun

fog — clouds that form on the surface of the ground

humid — a weather condition containing a large amount of moisture or water vapor

1. About half of it comes from plants, wet ground, rivers, and lakes. The other half of our precipitation on land is evaporated from the ocean.

2. Rain falls from clouds in the sky and runoff goes into bodies of water that eventually go to the oceans. Water in the ocean evaporates into the atmosphere, where it turns into precipitation. Some water soaks deep into the ground, creating the water table.

3. True

4. The age of the ocean is more like thousands of years rather than millions of years.

5. a. Cumulus; b. Stratus; c. Cirrus

6. Stratus and cumulus

Weather Chapter 3: – Water in the Atmosphere Worksheet 2

relative humidity — the amount of water vapor in the air compared to the amount of water vapor the air can contain at the point of saturation

stratus clouds — low-altitude gray clouds with a flat base

thunderstorm — a condition of weather that produces thunder, lightning, and rain

warm front — a boundary of warm air that is pushing out cold air in the atmosphere

water vapor — invisible water distributed throughout the atmosphere

1. A warm front is that part of the low-pressure system in which warmer air pushes the colder air back.

2. rising

3. "Red skies at night, sailor's delight; red sky in the morning, sailor's warning."

4. Both are low-pressure systems, but warm fronts replace cold air with warm air, while cold fronts replace warm air with cold air. Cold fronts move faster than warm fronts. Both fronts bring precipitation.

5. occluded

6. The most dreaded type of cold front is the Alberta Clipper in the central and eastern United States. This is an Arctic cold front in which cold Alaskan or Canadian air sweeps south or southeast into the United States as a low-pressure center passes to the east. Sometimes the weather can change from mild to below zero in a matter of several hours.

7. dew point

8. Most commonly, fog forms on clear nights when the temperature drops and the relative humidity rises to 100 percent, or the dew point. The air's water vapor condenses into liquid drops, forming fog.

9. Evaporation into the air moistens the air until fog forms.

10. Fog will occur when low clouds intersect the ground when trying to pass over high terrain. Sometimes fog will form even if moist air is uplifted over mountains.

11. inversion

Weather Chapter 4: Thunderstorms – Worksheet 1

> **downdraft** — a downward current of air
>
> **electricity** — a moving electric charge, such as in a thunderstorm
>
> **electrons** — a subatomic particle with a negative electrical charge

1. a. 242; b. 1,800; c. 40 to 50; d. 1.4; e. 20

2. False; cumulus clouds form thunderstorms

3. a. Large difference in temperature between the ground and the upper troposphere; b. Moisture in the lower atmosphere; c. A trigger to start the thunderstorm

4. moisture

5. moisture and extra heating

6. when the top of the cloud reaches about 25,000 feet

7. it weakens it

8. a. Sources of water for interior of mid-latitude continents; b. Regulates air temperature; c. Washes out and cleanses dust and pollutants from air particles; d. Maintains electrical balance; e. Places nitrogen into the soil

Weather Chapter 4: Thunderstorms – Worksheet 2

> **static electricity** — a build-up of electricity charge on an insulated body
>
> **updraft** — an upward current of air
>
> **latent heat** — the energy released or absorbed at constant temperature during a change in phase

1. Benjamin Franklin

2. a. If lightning approaches, seek shelter in a house, car, or low area, but not in a shed. Inside the house, do not use the telephone or any appliance. Do not bathe or take a shower.

   b. Stay away from water.

   c. Do not stand on a hilltop. Avoid being the tallest object.

   d. Do not seek shelter under an isolated tree.

   e. Stay away from metal pipes, fences, and wire clotheslines.

   f. If your hair stands on end while outside, immediately drop to the ground and curl into a ball

3. False

4. positively

5. 50,000°F

6. When lightning is near, it sounds like a sharp crack. But when it is farther away, it makes a rumbling noise.

7. Count the number of seconds between the lightning and the thunder and divide the seconds by five to determine the distance in miles.

Weather Chapter 5: Dangerous Thunderstorms – Worksheet 1

flash flood — a flood caused by a thunderstorm that deposits an unusual amount of rain on a particular area

hailstones — precipitation in the form of ice and hard snow pellets

supercell — a severe, well-organized thunderstorm with warm moist air spiraling upwards

tornado — a funnel-shaped column of air rotating up to 300 mph touching the ground

1. 100,000; one out of ten
2. True
3. A regular thunderstorm requires a strong updraft, but a severe thunderstorm requires a strong updraft and downdraft, formed when the ground is extra warm, the air is extra moist, and the air above is extra cool. The stronger the updraft, the more violent the storm.
4. Flash floods, mud slides
5. Strong wind; hail
6. Stronger updrafts in thunderstorms create larger sizes of hail.
7. The official world record is 8.0 inches (20 cm) in diameter that fell on Vivian, South Dakota, on July 23, 2010.
8. Weathermen call these chaotic winds microbursts or wind shear.

Weather Chapter 5: Dangerous Thunderstorms – Worksheet 2

dust devil — a relatively long-lived whirlwind on the ground formed on a clear, hot day

supercooled drops — drops of water that remain liquid below freezing

funnel cloud — a rotating whirlwind below a cloud that has not yet touched the ground

1. When it touches ground.
2. Waterspout. Waterspouts are most frequent in the Atlantic and Indian Oceans near the equator, in the Mediterranean Sea, and in the Gulf of Mexico. The largest number of waterspouts occurs in the Florida Keys of the United States.
3. Supercells are formed when factors that include strong speed, wind shear, and a height of at least 20,000 feet combine.
4. mechanism, life cycle
5. Scientists do not understand how strong tornadoes can form without a supercell. Moreover, many supercells do not produce tornadoes. Scientists still do not understand how stretching of the supercell downward produces a narrow tornado.
6. intensity, longevity
7. rain, lightning, southwest
8. wall
9. A tornado watch means conditions are favorable for a tornado to form. A tornado warning means a tornado or strong funnel cloud has been spotted.

Weather Chapter 5: Dangerous Thunderstorms – Worksheet 3

microburst — a small-scale intense downdraft from a thunderstorm

waterspout — a tornado over a body of water

tornado alley — the term used in the United States where tornadoes are more frequent, centered in northern Texas, Oklahoma, Kansas, and Nebraska

1. Great Plains, Rocky, Appalachian

2. Fujita

3. The most dangerous tornadoes are the thick black spiraling clouds that may be 2,000 feet (700 meters) across or more. These spin at up to 250 to 300 mph (400 to 480 kph) and are F4 or F5 tornadoes. They can sweep across the land at up to 50 mph (80 kph). Strong tornadoes can move along the ground for 100 miles (160 km) and have a damage path of over a mile (1.6 km) wide.

4. a. Go to the basement or the lowest floor of a house or building. Huddle close to the center of the house or building. Stay away from windows. Find a piece of strong furniture or a mattress to duck under or hide in a closet and wait until it is over.

   b. If you are in school when a tornado hits, an interior hallway on the lowest floor is safer than a classroom that has windows. Crouch near the wall. Bend over, placing your hands on the back of your head. By all means, stay out of auditoriums, gymnasiums, and other similar structures that have high ceilings.

   c. If you are in a mobile home or car, get out and seek shelter elsewhere.

   d. If you cannot find shelter, lie in a ditch or find the lowest, protected ground and cover your head with your hands

5. unusual

6. Doppler radar detects wind speed and direction inside thunderstorms up to 20 minutes before a tornado touches ground.

Weather Chapter 6: Hurricanes – Worksheet 1

Intertropical Convergence Zone — area near the equator where winds from different directions merge or mix

tropical depression — rainstorms with winds of 38 mph or less

monsoon — a seasonal reversing of the wind accompanied by corresponding changes in precipitation

1. a. Atlantic and Northeast Pacific hurricane; b. South Pacific and Indian Ocean cyclone; c. Northwest Pacific typhoon

2. a. One is the Intertropical Convergence Zone (ITCZ), an area where winds from different directions merge.

   b. The second rain process is the monsoon, in which certain areas of the tropics receive drenching rains for six months and are dry for the next six months.

3. the monsoon

4. a. Tropical depression: a rainstorm with 38 mph or less winds; b. Tropical storm: heavy rains with 39–74 mph winds; c. Hurricane: very heavy rains with wind speeds of 75 mph or greater

5. Hurricane hunters are pilots who fly into hurricanes and drop weather sensor into the storm that send back information about the hurricanes' characteristics.

6. 80

7. eyewall

8. weakens, falls apart

Weather Chapter 6: Hurricanes – Worksheet 2

storm surge — a coastal flood caused by rising ocean water during a storm, in particular, storms in the tropics, like hurricanes

tropical storm — a storm of heavy rain and winds between 30 and 74 mph

typhoon — another name for a hurricane

1. Six, two, four
2. decreased
3. increased
4. True
5. treasure
6. False
7. Rising ocean water rushing onto land causes most deaths during hurricanes.
8. Tornadoes, thunderstorms
9. a. Sandy in October 2012; b. Galveston, Texas, in 1900; c. Iniki in Hawaii in 1992; d. Katrina in the Gulf in August 2005; e. 2005; f. Bangladesh; g. Andrew on August 24, 1992; h. Bangladesh in 1991

Weather Chapter 7: Winter Storms – Worksheet 1

ice storm — a storm caused by rain falling into a lower atmosphere that is below freezing

sleet — precipitation that consists of frozen raindrops

wind chill factor — the temperature of windless air that would have the same cooling effect on exposed skin as a combination of wind speed and air temperature

blizzard — a very heavy snowstorm with violent winds

1. Melted snow
2. The earth tilts on its axis and orbits around the sun. As the earth rotates, its tilt causes some areas to experience shorter daylight hours, creating lower temperatures during the winter. As it continues to orbit the sun, the same areas begin to experience longer daylight hours, creating warmer temperatures during the summer.
3. less
4. around December 1
5. rain, snowstorms, Geography
6. Snow, Rain, snow
7. Snow's white color creates a high albedo, which means it reflects the sun, allowing the snow to melt slowly. This means the melted water soaks into the ground rather than causing floods.
8. 1846
9. Sutter's Fort near Sacramento, California
10. Sierra Nevada Mountains, October
11. Huge snowstorms hit the mountains.
12. 87, 47

Weather Chapter 7: Winter Storms – Worksheet 2

avalanche — a rapid flow of snow down a sloping surface

ice jam — the buildup of water caused by a blockage of ice

Northeaster — a storm that moves northeast along the east coast

1. The rain soaks deep into the ground to be used the next spring for growing crops. Winter rains add water to streams, rivers, springs, and wells.

2. Sometimes after heavy rains, the soil and their homes slide down the hill.

3. exciting, beautiful, dangers

4. 35 miles per hour

5. Answers will vary, but they should include examples involving extreme cold, avalanches, heavy snow, frostbite, hypothermia, etc.

6. March 12–15, 1993, on eastern North America

7. blizzard

8. once in a hundred years

9. 1888

10. Northeaster

11. tornadoes, thunderstorms, killed

12. storm surge

13. 56

14. 131

15. Hundreds of roofs collapsed, thousands of people were stranded, and millions were without electricity. For the first time in U.S. history, every major airport on the east coast was closed at one time or another. Interstate highways from Atlanta northward were closed, two ships sank, and the pounding surf destroyed houses along the coast. At least 270 people were killed, and 48 were missing at sea — three times the combined death toll from Hurricanes Andrew and Hugo. Property damage was estimated at $5 billion. It was the country's costliest winter storm ever.

Weather Chapter 7: Winter Storms – Worksheet 3

frostbite — localized damage to skin and tissues due to freezing

hypothermia — a condition in which the body temperature drops below that required for biological functions

inversion — a condition in which the atmospheric temperature increases upward

1. snow; sleet

2. dangerous, Frostbite, hypothermia

3. The most susceptible parts of the body are fingers, toes, ear lobes, and the tip of the nose.

4. Hypothermia happens when a person is out in the cold so long that the body temperature drops below normal. He or she starts shivering and is unable to stop. The person then becomes confused and forgets where he or she is. Speech becomes slurred, and the words don't make sense. Soon he or she becomes very tired and wants to sleep, anywhere.

5. True

6. 70, 25

7. Winter storms

8. The U.S. National Weather Service

Weather Chapter 8: Wild Weather – Worksheet 1

ball lightning — a glowing ball of red, orange, or yellow light found during a thunderstorm

chinook winds — foehn winds that are mild, gusty west winds found along the east slopes of the Rocky Mountains

Santa Ana winds — a foehn wind that blows westward from the mountains of southern California to the coast when a high pressure area settles over Nevada, Utah, and Idaho

St. Elmo's fire — a condition caused by a high charge of electricity

1. highest, 40 feet

2. rain forest, desert, hours

3. The western slopes

4. 102 feet, Mount Rainier

5. desert, 8 inches

6. The side toward the wind, or the windward side, receives rain, while the side away from the wind, the leeward side, receives very little rain.

Weather Chapter 8: Wild Weather – Worksheet 2

trade winds — the prevailing pattern of easterly winds in the tropics

foehn winds — a dry, warm downslope wind on the lee side of a mountain range

Rain forest — a woodland with a high annual rainfall and very tall trees and that is often found in tropical regions

lake-effect snowstorm — a snowstorm produced when cool air moves over long expanses of warm water, evaporating much moisture, and precipitating on downwind shores

1. fone

2. relatively warm and dry wind, descending down a mountain slope

3. They occur in almost all mountain regions of the mid and high latitudes, most frequently during the colder time of year.

4. cumulus, foehn wall

5. Cold, dry arctic air blowing over relatively warm water in the lakes causes high amounts of evaporation, which results in heavy snowfall when the air reaches the shore.

6. For long stretches of winter, chinook winds keep the area snow-free, enabling them to find dry grass in these areas.

7. They can fan grass fires out of control. Damaging winds, in some places peaking at 125 mph, can blow trucks and campers off the road and even derail trains. When arctic air retreats back up into Canada, the sparse water vapor from the chinook wind condenses as frost on the cold pavement, making it dangerous for drivers because, although the road looks dry, it's very slippery. Car windows quickly frost, causing

rapid condensation of water vapor on the windshield. All the windows turn white so the driver can't see the road.

8. Santa Ana winds occur in California. They create a funnel effect that picks up dust and fans wildfires. They occur at any time of the year.

9. St. Elmo's Fire is caused by a high charge of electricity in the air.

10. Buffalo, 2014, 7 feet

Weather Chapter 9: Climate in the Past – Worksheet 1

bogs — soft, waterlogged ground such as a marsh

environment — the surrounding circumstances or conditions around us

permafrost — permanently frozen subsoil found around polar regions

thermometer — an instrument used to indicate the temperature

1. Crocodiles, hippopotami, elephants, giraffes, crocodiles, and other animals. They also found fossils of fish and clams.

2. There are tens of thousands of rock art pictures painted by men. These pictures show many types of animals, people, and even whole villages that once existed in the Sahara Desert.

3. True

4. The scientific method requires observation. Because past weather conditions cannot be repeated or observed, we can only make educated guesses, which does not fulfill the scientific method.

5. A scientist must make an assumption about a past climate, create a model based on that assumption, and test how well data observed in the present fits the model.

6. Those who follow the creation-Genesis Flood model believe God was the only observer who created everything to reproduce after its own kind, and not into some other kind. In this model, most sedimentary rocks are a result of Noah's Flood. The evolution-uniformitarian model believes that every organism is a product of millions of years of evolution. This model eliminates Noah's Flood by stating that rocks form through a slow process of erosion and sedimentation.

7. They begin with two different models and beginning assumptions. They each believe in a different model, and so draw different conclusions.

Weather Chapter 9: Climate in the Past – Worksheet 2

Ice Age — a period of time marked by extensive glaciers on the face of the earth

Medieval Warm Period — a term used to refer to the warm period in the North Atlantic region, and possibly worldwide, from about A.D. 900 to 1200

uniformitarianism — the assumption that the same natural laws we observe today have always operated in the past, and is remembered by the saying "the present is the key to the past"

model — a conceptual representation of a phenomenon

1. sedimentary

2. The Genesis Flood

3. 1) A warmer atmosphere could have meant that the present-day cold climates were once warm climates. 2) The warm-climate animals could have been swept from high altitudes to the Polar Regions during the Flood.

4. The large number of warm-aspect fossils found all over the earth.

5. Coal

6. vegetation

7. Ice Age debris is found on top of Flood-deposited rocks and shows features that the Flood could not cause.

8. Volcanic ash blocked sunlight from entering the atmosphere, causing cooler temperatures. The volcanoes also warmed the water, creating more evaporation and greater amounts of snowfall in northern regions.

9. water vapor, cool

10. 700

Weather Chapter 9: Climate in the Past – Worksheet 3

sedimentary rock — type of rock formed by the deposition of sediments at the earth's surface in water, and later cemented

water vapor — invisible water distributed throughout the atmosphere

Little Ice Age — a term used to refer to the global cool period between about A.D. 1300 to 1880

1. 60

2. Genesis Flood, Ice Age

3. hundred

4. oceans, drier

5. present, inadequate

6. Around A.D. 800 to 1300

7. About 1300–1880

8. 1645–1715

Weather Chapter 10: Climate Change – Worksheet 1

El Niño — a warm current from the west that replaces the cool ocean current along Peru and Ecuador

fossil fuels — coal and oil derived from the remains of plant and animal organisms

ozone — a gas in the earth's upper atmosphere that screens most of the sun's harmful ultraviolet radiation

1. climate

2. change, constant

3. cooler

4. It flows northward along the South American coast of Peru and Ecuador and the warming normally occurs around Christmas.

5. Every two to seven years, the warm current that flows northward along the South American coast becomes warmer than normal. It is poor in nutrients, resulting in few fish. It brings heavy rains and flooding to Peru and Ecuador. It causes dry weather in other areas like India and stronger winds in North America with milder winters.

6. simultaneously

7. North Atlantic, Pacific Decadal

8. False

9. It results from a change in air pressure between Iceland and the Azores Islands. These pressure fluctuations have a control on the direction and strength of the general west winds and storm tracks.

10. True

Weather Chapter 10: Climate Change – Worksheet 2

plankton — tiny plant and animal organisms found in the oceans

pollution — harmful or unsafe waste products

greenhouse (effect) warming — the phenomena of a steady, gradual rise of temperatures

1. True

2. 1 Thessalonians 5:21: "Examine everything carefully; hold fast to that which is good" (NASB).

3. False

4. False

5. Answers will vary, but possibilities include the heat island effect, changing instrument shelter location, placing shelters near heat sources, and changing instruments.

6. Satellite images

Weather Chapter 10: Climate Change – Worksheet 3

ultraviolet light — the range of wavelengths just beyond violet in the visible spectrum; invisible to humans, yet capable of causing skin cancer

climate — the weather conditions that are particular to a certain area, such as wind, precipitation, and temperature

La Niña — an ocean/atmospheric phenomenon that is the opposite of El Niño

heat-island effect — the warming effect in cities, compared to rural areas, caused by human activities

1. Hurricane landfalls, droughts, floods, tornadoes or thunderstorms, East Coast winter storms, heat waves, and cold spells.

2. Should include one of the following: increased pollution that causes health problems; smoke and other pollutants from cars and factories that can cause acid rain; the addition of compounds of the element chlorine into the air that have caused chemical changes in the stratosphere, namely the loss of ozone.

3. changes, economic

4. Ozone

5. The ozone layer protects the earth from harmful ultraviolet rays. Without the ozone layer, we would all die. The little bit of ultraviolet light that does make it through the atmosphere kills excess bacteria and produces vitamin D in our skin.

6. scientific, global warming

7. chlorofluorocarbons

8. complex

9. ozone

10. with oxygen, without oxygen

Weather Chapter 11: God, Creation and You – Worksheet 1

   creation — the formation of everything by God

   environmentalist — someone concerned with the environment

   Theory of Evolution — a theory that claims that life came from non-life, new species arise through natural selection over millions of years.

1. Genesis 2:15: "The LORD God took the man and put him into the garden of Eden to cultivate it and keep it" (NASB).

2. False

3. Stewards

4. False

5. Answers will vary, but they should include sin as the root of the pollution problem and that there is little hope for solving environmental problems without Jesus. Christians should work to solve the pollution of souls as well as pollution of the earth.

Weather Chapter 11: God, Creation and You – Worksheet 2

   Pantheist — the idea that the universe or nature is divine or everything is god

   steward — a person who looks out favorably for the property of another

1. weather vane, stationary, pointer

2. Mercury

3. rain gauge

4. Barometers

5. the force from the weight of the air in the atmosphere exerted upon the earth's surface

6. clearer

The New Astronomy Book ➼ Worksheet Answer Keys

Introduction to New Astronomy Book: What is Astronomy? – Worksheet 1

astronomical – part of or related to aspects within astronomy; also used to describe extremely large distances or amounts

comet – small bodies in space that contain frozen dust, gases, and even rocks that have orbits of their own; made up of a nucleus, often with a trail of particles and dust that follows it

astrophysics – the application of modern physics to the study of astronomy

astronomy – the study of heavenly bodies, things outside of the earth, including the sun, moon, and stars

spectroscopy – the study of spectra

1. To separate night and day; signs to mark sacred times, days, and years; to give light on the earth
2. Sun for the day, moon for the night
3. Verses 14–15
4. As being good
5. Day three

New Astronomy Chapter 1: The Night Sky – Worksheet 1

constellations – a group of stars that seems to form a pattern or shape

axis – an imaginary line, vertical and horizontal, around which a planet or other body rotates

celestial – a reference term related to the universe and objects within it

circumpolar – means "around the pole," referring to stars that from a given location neither rise nor set but appear to circle the pole

revolution – circular motion around another body

pagan – refers to ancient cultures that were not based on Christianity, Islam, or Judaism

retrograde motion – when a planet appears to move east to west with respect to the stars, opposite from its normal motion

1. They don't rise or set, but instead move in a counterclockwise circle.
2. They appear to move clockwise around the south celestial pole; no, it does not.
3. Winter; summer
4. A NEO is a near earth object. NEOs include comets and asteroids close to earth; answers will vary.
5. No. Study of them can be traced back to ancient civilizations like Egypt, Greece, and Babylon.
6. The moon
7. A year; one month
8. Rotation
9. Mercury, Venus. Mars, Jupiter, and Saturn
10. Uranus and Neptune

New Astronomy Chapter 2: The Moon – Worksheet 1

unmanned – refers to missions or spacecraft that do not have humans onboard to operate them in space

maria – Latin for "seas"; refers to the darker areas of the moon's surface, at lower elevations; so called because astronomers once thought they might be bodies of water

highlands – areas of the craters on the surface of the moon that appear lighter and are at a high elevation

lunar – referring to features and aspects related to and upon the moon

impact basins – refers to large, round, dark features on the surface of the moon, created by impacts of large space objects of some kind

asteroids – considered to be minor planets, especially if located in the inner region of our solar system

bombardment – when an object in space, like the moon, has been struck by many other objects (comets, meteors, etc.) in space

1. One-fourth the earth's diameter and containing only a little more than 1 percent of the mass of the earth.

2. 240,000 miles

3. Mercury, Gemini, and Apollo

4. The first manned lunar landing of Apollo 11

5. When an object in space rotates on its axis at the same rate that it revolves around another body, like a planet; the moon

6. Galileo

7. Volcanic activity or impacts

8. This is because some portions of the moon reflect light better than other portions. The lighter portions of the moon are made of rock similar to granite, while the darker areas are made of rock similar to basalt. Granite and basalt are common rocks on earth. Granite usually is lighter than basalt.

9. They are periods of bombardment thought to have occurred over a long period of time to create the features of the moon that we see today.

10. It doesn't. It reflects light it gets from the sun.

New Astronomy Chapter 2: The Moon – Worksheet 2

The sun also produces tides, but because it is so much farther away than the moon, its tides aren't nearly as high. At new and full moon, both lunar and solar tides work together, and we say that this is spring tide. This name refers to how much the tides leap from very low levels at low tide to very high levels at high tide. On the other hand, at the quarter phases, the lunar and solar tides compete. This is a neap tide, and low tide is large, but the difference between high and low tide is small at neap tide.

1. The earth's gravity pulls on the moon, causing the moon to orbit the earth. But the moon's gravity also pulls on the earth and even alters its shape a little. The moon's gravity also impacts the ocean tides.

2. No. The moon's gravity pulls on the side of the earth facing the moon (sublunar) more than it pulls on the earth's center. And the moon's gravity pulls on the earth's center more than it pulls on the side of the earth away from the moon (antipodal). This differential force of the moon's gravity stretches the earth.

3. This is when the earth's shadow falls on the moon.

4. They can only happen when the moon is full, and the earth's shadow actually falls on the moon, not above or below as it normally does.

5. Umbra

6. They knew the earth's shadow caused lunar eclipses, and that it was always circular. They also realized that only a sphere casts a circular shadow.

7. No, this is a myth.

8. The corona is the outermost layer of the sun; prominences are loops of gas that follow the sun's magnetic fields.

9. The sun is 400 times larger, but it is also 400 times farther away, giving the illusion of it being the same size as the moon.

10. An ellipse

Activity: Answers will vary, but they need to show that orbiting cycles of different celestial bodies aided in determining the passage of time. They should also mention that marking time is one of the purposes given for these objects in the Bible.

New Astronomy Chapter 3: The Solar System – Worksheet 1

satellites – another word to describe moons that orbit planets

planet – a large celestial body that is orbiting around a star

minor planets – another name preferred by astronomers and given to bodies orbiting the sun that are also called asteroids, smaller than planets

nucleus – the center portion of a comet, usually made of ice and dust

ellipse – shaped like a flattened circle, referring to the shape of some galaxies that do not have disks or spiral arms

coma – the gas and dust that expand to form a large cloud when a comet passes close to the sun

orbital period – the length of time required for an orbiting body to complete one orbit

1. Mercury

2. In 2006, astronomers decided it was too small to be a planet, calling it instead a minor planet.

3. Pluto, Eris, Ceres, Haumea, Makemake

4. Mercury, Mars, Earth, Venus, Jupiter

5. Saturn, Uranus, Neptune, Pluto

6. Nucleus; ice with a little rocky material

7. International Astronomical Union

8. Comet Shoemaker-Level IX

9. Wearing out, collision with a planet, or being ejected from the solar system

10. They orbit a planet rather than the sun.

New Astronomy Chapter 4: Two Kinds of Planets – Worksheet 1

Astronomical unit – a unit of distance, AU, which is based on the average distance between the earth and sun

Mass – the measurement of how much matter a body (planet, star, etc.) may have

Density – measure of how closely packed matter is, expressed in grams/cubic centimeter

Rotation periods – The period of time that it takes an object to spin on its axis once

1. Because distances in space like miles and kilometers don't work well across vast areas; they would be enormous and impractical to use.

2. Mercury

3. Saturn

4. Venus and Uranus

5. Earth

6. An impact would have altered their orbits, and they are among the most circular, making an impact improbable in explaining their clockwise rotation.

7. terrestrial and jovian

8. terrestrial

9. Jovian

10. It is so close to the sun that we cannot see it in the night sky, and the atmosphere on earth gets in the way of us seeing it because it is positioned so low in the sky.

New Astronomy Chapter 4: Two Kinds of Planets – Worksheet 2

terrestrial – "earth-like"; also refers to the four closest planets to our sun

jovian – Jupiter-like, refers to planets with certain characteristics — farther from the sun, more mass, and much larger

Galilean satellites – refers to the four largest moons of Jupiter, named for the person who discovered them

cryovolcanism – volcanic activity that takes place on very cold objects in space

1. Venus

2. The clouds — they thought they were made of water droplets like they are on earth.

3. It traps the infrared radiation rather than letting it radiate away.

4. Mars

5. Channels appear to signal it once had liquid water; it used to have a more extensive atmosphere, making the planet warmer and wetter.

6. Jupiter

7. It is the only solid-surface body without impact craters.

8. Many scientists think there is a layer of liquid water between the thick ice layer on the surface and the rocky core at its center.

9. The gravitational pull of moons, like those of Saturn, tear rings apart — making their chance of surviving millions or billions of years after being formed very unlikely.

10. Uranus

New Astronomy Chapter 5: The Sun – Worksheet 1

fission – a process where a large atom, such as uranium, fissions or breaks into smaller atoms, releasing a lot of energy

fusion – smaller atoms combining into larger atoms, creating energy

potential energy – the energy that an object has because of its location

paradox – an idea that seems contradictory, so requires an explanation

sunspots – areas that appear to be dark spots on the photosphere of the sun

photosphere – refers to what appears to be the surface area of the sun

umbra – a shadow of an astronomical body in which the light source has been completely blocked out, such as occurs during a total eclipse

penumbra – the partial shadow of the earth on the moon

magnetic polarity – refers to having a magnetic north or south, as sunspots have

1. No one knows, but scientists think it is nuclear powered and operated by fusion.
2. Fission
3. It wouldn't last long enough for the sun to still be shining after tens of millions of years.
4. No; they think it has stayed closed to current temperatures.
5. The earth would be too hot for life to exist.
6. The question of how the earth and sun could be billions of year old yet have maintained the same average temperature.
7. Photosphere
8. By noting changes in the sunspots from day to day and how they move during the sun's rotation.
9. 11 years
10. They are regions of magnetic fields, and the sunspots in the pair have opposite polarity.

New Astronomy Chapter 6: Telescopes – Worksheet 1

spectrograph – a device or program that divides light up into its wavelengths, or colors

wavelengths – refers to how long a light wave is

electromagnetic radiation – the name for invisible wavelengths of light

infrared radiation – also known as heat radiation, refers to the radiation that has wavelengths longer than what is visible to the human eye

microwaves – a part of the electromagnetic spectrum beyond infrared but before the radio part of the spectrum

radio waves – the longest wavelength electromagnetic radiation

spectroscopy – the study of spectra

spectral lines – unique patterns of absorption or emission lines that help scientists determine the composition of astronomical bodies in space

Doppler Effect – the shift in spectral lines from their normal wavelengths that help to show whether an object in space is moving toward or away from us

1. Refractors and reflectors
2. By the diameter of its lens
3. Things in space farther away from us appear very faint, and larger telescopes can collect more light to be able to see them.
4. So they measure the brightness of stars or the structure of astronomical bodies.
5. No, its changes.

6. red, orange, yellow, green, blue, indigo, violet

7. Ultraviolet

8. It blocks most of these.

9. They pass through the atmosphere.

10. Different elements form different spectral lines so they can be used to tell the composition of the body.

New Astronomy Chapter 7: History of Astronomy – Worksheet 1

parallax – the visual shift slightly back and forth each year of the stars, based on where we are viewing it from either side of earth's orbit

Geocentric Theory – means "earth-centered," or the idea that the earth remained motionless while the sun orbited it

Heliocentric Theory – means "sun-centered," or the theory that planets of our solar system orbit around the sun

epicycle – a smaller circle that moved on a larger circle that Ptolemy used to explain retrograde motion of planets

retrograde motion – when a planet appears to move east to west with respect to the stars, opposite its normal motion

laws of planetary motion – three laws that describe the motions of the planets discovered by Johannes Kepler

physics – the study of motion, energy, matter, and force

1. By looking at the moon during a lunar eclipse; the earth's shadow was always a circle; and watching how high in the sky the stars appear when traveling north and south

2. Eratosthenes; by measuring how high the sun was in the sky on the same date from two locations running along the north and south line. Then with the distance between the locations, the difference in the height of the sun in the two locations, he calculated the circumference of the earth.

3. Whether the sun moved around the earth or the earth moved around the sun

4. They could not see parallax.

5. It did not fit the idea that the heavens were perfect and followed perfect rules.

6. He created epicycles to try and adjust for these imperfections.

7. Nicolaus Copernicus

8. The phases of Venus, craters on the moon, and four moons orbiting Jupiter

9. Kepler

10. President Kennedy

New Astronomy Chapter 8: Stars – Worksheet 1

light years – the distance that light travels in a year; used to express distances between stars and other space objects

absolute magnitude – how bright a star would be if it were at the standard distance of 10 parsecs

parsec – a unit of measure of distance that astronomers use that is equal to 3.26 light years

binary systems – binary, meaning, two; a star system in which two stars are orbiting one another

white dwarf stars – stars that are a lot smaller than the sun, about the size of the earth, but they are so hot, they are white

neutron star – a very dense star that is about the size of the earth, but has several times the mass of the sun

black hole – a star that is so massive and small that its gravity is very strong, so strong that light cannot escape

supernovae – the very bright explosions of certain stars that astronomers think are the end of such stars

pulsar – another name for neutron stars that appear to pulse because of flashes of radiation seen

1. Trigonometry
2. Magnitude
3. a. -30; b. -12; c. 2
4. How bright a star appears to be to us
5. Its temperature
6. Hydrogen
7. How much mass stars have
8. They are less dense than the air we breathe.
9. The effect of its gravity on nearby objects
10. No, it is temporary

New Astronomy Chapter 9: Extrasolar Planets – Worksheet 1

extrasolar – planets that orbit a star beyond our solar system

exoplanet – another name for extrasolar planets, or planets that are believed to be orbiting a star outside of our solar system

main sequence star – a grouping of stars rated as normal like the sun; can be as small as 1/10th the size of the sun or up to ten times larger than the sun

transit – when a smaller body passes in front of a larger body; a type of eclipse with a much smaller body blocks out only a small portion of a bigger body's light

habitable zone – the theoretical range of distance between a star and an orbiting planet that might have liquid water on the surface

1. No, they are very common
2. They were too far away and too close to their own stars for their light to be seen.
3. By adding a lowercase letter to the name of the parent star
4. Look for motion in the star as the planet orbits it
5. Less
6. Stars
7. By decreasing the amount of the brightness we see
8. Kepler
9. They are searching for other inhabited planets like earth that they feel evolved naturally.
10. No other planets with life have been found as on earth; theories on how the universe formed are being proven to not be what they thought; how planets can be orbiting a pulsar

New Astronomy Chapter 10: Star Clusters and Nebulae – Worksheet 1

star clusters – groups of stars spaced much closer together than surrounding stars and held together by their gravity

open clusters – one of two types of star clusters, usually with fewer stars and an irregular appearance

globular clusters – a type of star cluster that usually is spherical, with many, many stars

interstellar medium – thin gas and other small materials that exist between stars

silicate – a type of rock

nebulae – what are thought to be clouds of glowing gas in space

dark nebula – a dense dust cloud in space that blocks the light of stars and so appears dark

reflection nebula – a nebula where the light of hot, bright stars near the nebula scatters, or reflects off the dust and appears blue in images

molecular clouds – the largest and densest regions of the interstellar medium, containing molecules as well as atoms and dust particles

1. Stars that are distributed in space but are not in closely packed star clusters

2. Yes

3. Three

4. Open and globular

5. Globular

6. Open

7. Thin, gas, clumpy, and includes dust

8. Nebula

9. Electrons are removed from the atoms

10. No — theoretically it would be too slow to be seen, and there is a question of whether stars are still being born.

New Astronomy Chapter 11: Our Galaxy: The Milky Way – Worksheet 1

spiral tracers – objects that lie along spiral arms of galaxies, and so can be used to trace the locations of spiral arms

nebulae – what are thought to be clouds of glowing gas in space

spiral nebulae – what astronomers used to call what are now known as spiral galaxies

Cepheid variables – pulsating giant and supergiant stars that follow a period-luminosity relationship, making them useful for measuring distances

spiral galaxies – two types of galaxies that have the characteristic of spiral arms

irregular galaxies – galaxies that cannot be placed into the common categories of galaxies like spiral or elliptical

Local Group – the group of galaxies that includes the Milky Way and another spiral galaxy, M31

dwarf galaxies – very small galaxies, many of which orbit much larger galaxies as a satellite

1. Galaxy

2. Herschel

3. About 100,000 light years

4. Barred spiral galaxy

5. Based on mass and rotation speeds, the arms would have long ago "smeared out."

6. The halo

7. The telescope

8. They were solar systems in development.

9. No, not usually

10. Virgo Cluster

New Astronomy Chapter 12: Light-Travel-Time Problems – Worksheet 1

general relativity – the current theory of gravity, and how it relates space and time to matter and energy

white hole theory – The theory based on the idea that matter and energy fly outward from a white hole. As matter and energy exit a white hole, the white hole would get smaller and eventually go away. However, time would pass at very different rates inside and outside the white hole. If the earth is inside the white hole and stars are outside, millions of years might pass for the stars and the space outside, but only a few days might have passed on earth.

dimension – a fundamental quantity of space or time, usually associated with a direction and measured in meters or seconds

ASC solution – stands for the anisotropic synchrony convention solution by Dr. Lisle to the problem of light-travel-time

dasha solution – a solution to the light-travel-time problem that suggests that God miraculously and quickly brought the light of distant stars to the earth on day 4 of the creation week

Ex Nihilo – Latin word that means "out of nothing"; can refer to the creation, when God created everything from nothing

1. How we can see any galaxies if they are millions or billions of light years away if the universe is only 6,000 years old.

2. Can include any of the following: perhaps God made the universe fully mature and fully functioning, or the speed of light was much higher in the past, the white hole theory, the ASC solution, or the dasha solution.

3. Choices include Danny Faulkner, John Hartnett, Russ Humphreys, and Jason Lisle.

4. Light is assumed to travel the same speed during the two-way trip (there and back).

5. The plants would need to be either mature instantly or grow mature very quickly in a couple of days.

6. Answers will vary per student.

List the three issues we must overcome to allow humans to visit distant places in the universe.

7. going long distances in a short period of time

8. manning spacecraft for long voyages, taking into account the average human life span

9. vehicle design that provides food, shelter, care, water, and scientific study equipment

10. Not really — creating the entire universe is something bigger than we can really comprehend, and scientists, both secular and creationist, are still discovering remarkable aspects of it!

New Astronomy Chapter 13: The Expanding Universe – Worksheet 1

Hubble relation – concept that redshifts and the distances of galaxies are related, such as the greater the redshift of a galaxy, the greater the distance of the galaxy

redshifted – when the Doppler shift moves light to longer wavelengths (toward red), indicating objects are moving away from us

blueshifted – when the Doppler shift moves light to shorter wavelengths (toward blue), indicating objects are moving toward us

extra-galactic astronomy – the study of galaxies other than the Milky Way

standard candle – an object for which we think we know the actual brightness, or absolute magnitude, that can be used to determine distances

1. Edwin Hubble in 1929

2. How fast or slow an object or source of light is moving toward or away from us

3. Most of what were thought to be nebulae of the time were redshifted.

4. The redshifted galaxy is farther away

5. Not really

6. No, it is just one of several theories about the universe; creation science even has a number of theories that could account for this.

7. The vast distances between objects

8. No, many are being commercially developed.

9. No, not yet.

10. No

New Astronomy Chapter 14: Quasars and Active Galaxies – Worksheet 1

radio stars – the name used for quasars before their nature was better known, so called because they looked like stars but gave off radio emissions

quasi-stellar objects – another name for quasars

quasars – small, high redshift objects that likely are very far away and very bright

synchrotron radiation – a special form of radiation created by very strong magnetic fields interacting with fast-moving charged particles

synchrotron spectra – a unique kind of spectrum produced by sources of synchrotron radiation

1. 1950s

2. Blue

3. Quasi-stellar objects

4. A trillion

5. No

6. They were thought to be supermassive black holes.

7. Evidence for it

8. 2.6 million times more

9. It is not as bright as those being fed large amounts of matter.

10. How thick the disk around the black hole may be, and how narrow are the jets that eject matter perpendicular to the disk; their orientation to the disk and jets to our line of sight

Activity: Names of constellations

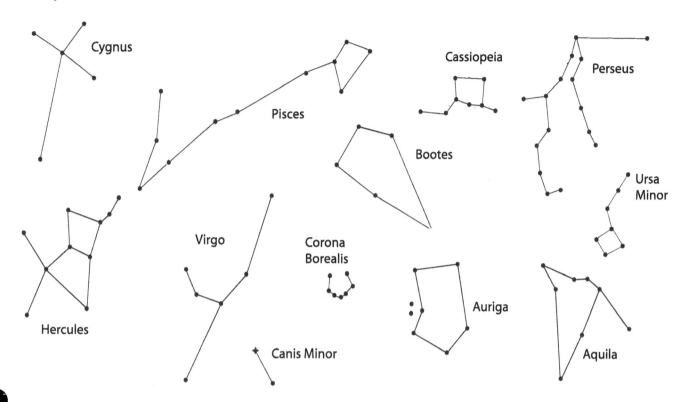

New Astronomy Chapter 15 & Conclusion: Cosmology – Worksheet 1

cosmology – the study of the structure of the universe

filaments – lines and flat or curved surfaces where galaxies seem to cluster

voids – areas between clusters of galaxies that appear to be open or empty

steady-state theory – the idea that matter popped into existence to keep a constant density in the universe

continuous-creation theory – alternate name for the steady-state theory

big-bang theory – a popular but unproven idea that the universe began about 13.8 billion years ago in a very hot, dense state and has been expanding and cooling since

cosmic microwave background – radiation of a particular type in the microwave part of the spectrum that is coming from all directions

horizon problem – a problem with the big-bang model in that the cosmic microwave background has the same temperature in every direction

flatness problem – a problem in the big-bang model that refers to the certain type of geometry that the universe has

1. Yes, they are.

2. A physicist or astronomer who specializes in the study of cosmology or structure of the universe

3. The big-bang theory relies on the existence of CMB, and when evidence of CMB was found, the steady-state theory was discounted.

4. The earth being formed before the rest of the universe, including stars and other planets, and the biblical timeline of creation as having happened 6,000 years ago

5. The horizon problem and the flatness problem

6. Scientists changed the big-bang model to fit the data.

7. A larger expansion rate, inclusion of inflation and dark matter, dark energy, and the string theory

8. There is no evolution-based alternative to it for secular scientists.

9. The Creation Museum

10. Project Moon-Blink

The New Weather Book ➝ Unit Quiz Answer Key

Unit One Quiz, chapters 1-3

1. **climate** — the weather conditions that are particular to a certain area, such as wind, precipitation, and temperature

2. **nitrogen** — a naturally occurring element that is responsible for around four-fifths of the earth's atmosphere

3. **oxygen** — a colorless, odorless gas that is 21 percent of our atmosphere; essential for plant and animal respiration

4. **precipitation** — falling moisture in the form of rain, sleet, snow, hail, or drizzle

5. **cirrus clouds** — a high altitude cloud made of ice crystals that appears thin, white, and feathery

6. **cold front** — a boundary of cold air, usually moving from the north or west, which is displacing the warm air

7. **evaporation** — to change into a vapor such as the evaporation of water by the warming of the sun

8. **stratus clouds** — low-altitude gray clouds with a flat base

9. **warm front** — a boundary of warm air that is pushing out cold air in the atmosphere

10. **water vapor** — invisible water distributed throughout the atmosphere

1. When Adam and Eve disobeyed God's commands, they allowed evil to enter the world. Bad weather is a result of the presence of sin in the world.

2. God created the world with a perfect design and order, which allows us to predict hours of daylight, seasons, and weather.

3. The sun is responsible for differences in heating around the earth. At night, infrared radiation cools the earth, and the sunrise warms the earth during the day.

4. Cold front: line with triangles. Warm front: line with semicircles. Stationary front: line with triangles on top and semicircles on the bottom. Occluded front: line with alternating triangles and semicircles on top.

5. Meteorologists can only interpret weather maps created by computers, which are not perfect. Despite all the tools they have, meteorologists do not have a complete understanding of the earth and its atmosphere.

6. The general circulation is the average flow of air in various locations. The earth has six average circulations, which determines the climate in different areas.

7. Rain falls from clouds in the sky and runoff goes into bodies of water that eventually go to the oceans. Water in the ocean evaporates into the atmosphere, where it turns into precipitation. Some water soaks deep into the ground, creating the water table.

8. Stratus and cumulus

9. Both are low-pressure systems, but warm fronts replace cold air with warm air, while cold fronts replace warm air with cold air. Cold fronts move faster than warm fronts. Both fronts bring precipitation.

10. Most commonly, fog forms on clear nights when the temperature drops and the relative humidity rises to 100 percent, or the dew point. The air's water vapor condenses into liquid drops, forming fog.

Unit Two Quiz, chapters 4-6

1. **electrons** — a subatomic particle with a negative electrical charge

2. **static electricity** — a build-up of electricity charge on an insulated body

3. **latent heat** — the energy released or absorbed at constant temperature during a change in phase

4. **supercell** — a severe, well-organized thunderstorm with warm, moist air spiraling upwards

5. **tornado** — a funnel-shaped column of air rotating up to 300 mph touching the ground

6. **supercooled drops** — drops of water that remain liquid below freezing

7. **funnel cloud** — a rotating whirlwind below a cloud that has not yet touched the ground

8. **microburst** — a small-scale intense downdraft from a thunderstorm

9. **monsoon** — a seasonal reversing of the wind accompanied by corresponding changes in precipitation

10. **tropical storm** — a storm of heavy rain and winds between 30 and 74 mph

1. 50,000°F

2. Count the number of seconds between the lightning and the thunder and divide the seconds by five to determine the distance in miles.

3. A regular thunderstorm requires a strong updraft, but a severe thunderstorm requires a strong updraft and downdraft, formed when the ground is extra warm, the air is extra moist, and the air above is extra cool. The stronger the updraft, the more violent the storm.

4. Stronger updrafts in thunderstorms create larger sizes of hail.

5. When it touches ground.

6. A tornado watch means conditions are favorable for a tornado to form. A tornado warning means a tornado or strong funnel cloud has been spotted.

7. Fujita

8. Doppler radar detects wind speed and direction inside thunderstorms up to 20 minutes before a tornado touches ground.

9. Hurricane hunters are pilots who fly into hurricanes and drop weather sensors into the storm that send back information about the hurricanes' characteristics.

10. Rising ocean water rushing onto land causes most deaths during hurricanes.

Unit Three Quiz, chapters 7-9

1. **ice storm** — a storm caused by rain falling into a lower atmosphere that is below freezing

2. **blizzard** — a very heavy snowstorm with violent winds

3. **hypothermia** — a condition in which the body temperature drops below that required for biological functions

4. **inversion** — a condition in which the atmospheric temperature increases upward

5. **ball lightning** — a glowing ball of red, orange, or yellow light found during a thunderstorm

6. **St. Elmo's fire** — a condition caused by a high charge of electricity

7. **trade winds** — the prevailing pattern of easterly winds in the tropics

8. **foehn winds** — a dry, warm downslope wind on the lee side of a mountain range

9. **environment** — the surrounding circumstances or conditions around us

10. **thermometer** — an instrument used to indicate the temperature

1. Snow's white color creates a high albedo, which means it reflects the sun, allowing the snow to melt slowly. This means the melted water soaks into the ground rather than causing floods.

2. Answers will vary, but they should include examples involving extreme cold, avalanches, heavy snow, frostbite, hypothermia, etc.

3. The side toward the wind, or the windward side, receives rain, while the side away from the wind, the leeward side, receives very little rain.

4. Areas in the United States along the east slopes of the Rocky Mountains from Alberta, Canada, to New Mexico.

5. St. Elmo's Fire is a high charge of electricity in the air that causes pointed objects to glow when thunderstorms are in the area.

6. Cold, dry arctic air blowing over relatively warm water in the lakes causes high amounts of evaporation, which results in heavy snowfall when the air reaches the shore.

7. The scientific method requires observation. Because past weather conditions cannot be repeated or observed, we can only make educated guesses, which does not fulfill the scientific method.

8. A scientist must make an assumption about a past climate, create a model based on that assumption, and test how well data observed in the present fits the model.

9. 1) A warmer atmosphere could have meant that the present-day cold climates were once warm climates. 2) The warm-climate animals could have been swept from high altitudes to the Polar regions during the Flood.

10. Volcanic ash blocked sunlight from entering the atmosphere, causing cooler temperatures. The volcanoes also warmed the water, creating more evaporation and greater amounts of snowfall in northern regions.

Unit Four Quiz, chapters 10-11

1. **El Niño** — a warm current from the west that replaces the cool ocean current along Peru and Ecuador

2. **ozone** — a gas in the earth's upper atmosphere that screens most of the sun's harmful ultraviolet radiation

3. **plankton** — tiny plant and animal organisms found in the oceans

4. **pollution** — harmful or unsafe waste products

5. **greenhouse (effect) warming** — the phenomena of a steady, gradual rise of temperatures

6. **ultraviolet light** — the range of wavelengths just beyond violet in the visible spectrum. Invisible to humans, yet capable of causing skin cancer

7. **La Niña** — an ocean/atmospheric phenomenon that is the opposite of El Niño

8. **creation** — the formation of everything by God

9. **environmentalist** — someone concerned with the environment

10. **steward** — a person who looks out favorably for the property of another

1. Every two to seven years, the warm current that flows northward along the South American coast becomes warmer than normal. It is poor in nutrients, resulting in few fish. It brings heavy rains and flooding to Peru and Ecuador. It causes dry weather in other areas like India and stronger winds in North America with milder winters.

2. 1 Thessalonians 5:21: "Examine everything carefully; hold fast to that which is good" (NASB).

3. Answers will vary, but possibilities include the heat island effect, changing instrument shelter location, placing shelters near heat sources, and changing instruments.

4. Satellite images

5. Hurricane landfalls, droughts, floods, tornadoes or thunderstorms, East Coast winter storms, heat waves, and cold spells.

6. chlorofluorocarbons

7. Genesis 2:15: "The LORD God took the man and put him into the garden of Eden to cultivate it and keep it" (NASB).

8. stewards

9. False

10. Answers will vary, but they should include sin as the root of the pollution problem and that there is little hope for solving environmental problems without Jesus. Christians should work to solve the pollution of souls as well as pollution of the earth.

The New Weather Book ━➤ Test Answer Key

1. nitrogen — a naturally occurring element that is responsible for around four-fifths of the earth's atmosphere

2. oxygen — a colorless, odorless gas that is 21 percent of our atmosphere; essential for plant and animal respiration

3. cirrus clouds — a high altitude cloud made of ice crystals that appears thin, white, and feathery

4. evaporation — to change into a vapor such as the evaporation of water by the warming of the sun

5. stratus clouds — low-altitude gray clouds with a flat base

6. static electricity — a build-up of electricity charge on an insulated body

7. supercell — a severe, well-organized thunderstorm with warm, moist air spiraling upwards

8. tornado — a funnel-shaped column of air rotating up to 300 mph touching the ground

9. supercooled drops — drops of water that remain liquid below freezing

10. tropical storm — a storm of heavy rain and winds between 30 and 74 mph

11. ice storm — a storm caused by rain falling into a lower atmosphere that is below freezing

12. blizzard — a very heavy snowstorm with violent winds

13. hypothermia — a condition in which the body temperature drops below that required for biological functions

14. St. Elmo's fire — a condition caused by a high charge of electricity

15. trade winds — the prevailing pattern of easterly winds in the tropics

16. foehn winds — a dry, warm downslope wind on the lee side of a mountain range

17. environment — the surrounding circumstances or conditions around us

18. thermometer — an instrument used to indicate the temperature

19. El Niño — a warm current from the west that replaces the cool ocean current along Peru and Ecuador

20. ozone — a gas in the earth's upper atmosphere that screens most of the sun's harmful ultraviolet radiation

21. greenhouse (effect) warming — the phenomena of a steady, gradual rise of temperatures

22. ultraviolet light — the range of wavelengths just beyond violet in the visible spectrum; invisible to humans, yet capable of causing skin cancer

23. La Niña — an ocean/atmospheric phenomenon that is the opposite of El Niño

24. creation — the formation of everything by God

25. steward — a person who looks out favorably for the property of another

1. When Adam and Eve disobeyed God's commands, they allowed evil to enter the world. Bad weather is a result of the presence of sin in the world.

2. The sun is responsible for differences in heating around the earth. At night, infrared radiation cools the earth, and the sunrise warms the earth during the day.

3. The general circulation is the average flow of air in various locations. The earth has six average circulations, which determines the climate in different areas.

4. Rain falls from clouds in the sky, and runoff goes into bodies of water that eventually go to the oceans. Water in the ocean evaporates into the atmosphere, where it turns into precipitation. Some water soaks deep into the ground, creating the water table.

5. Count the number of seconds between the lightning and the thunder and divide the seconds by five to determine the distance in miles.

6. A tornado watch means conditions are favorable for a tornado to form. A tornado warning means a tornado or strong funnel cloud has been spotted.

7. Doppler radar detects wind speed and direction inside thunderstorms up to 20 minutes before a tornado touches ground.

8. Hurricane hunters are pilots who fly into hurricanes and drop weather sensors into the storm that send back information about the hurricanes' characteristics.

9. Snow's white color creates a high albedo, which means it reflects the sun, allowing the snow to melt slowly. This means the melted water soaks into the ground rather than causing floods.

10. Answers will vary, but they should include examples involving extreme cold, avalanches, heavy snow, frostbite, hypothermia, etc.

11. The side toward the wind, or the windward side, receives rain, while the side away from the wind, the leeward side, receives very little rain.

12. Areas in the United States along the east slopes of the Rocky Mountains from Alberta, Canada, to New Mexico.

13. St. Elmo's Fire is a high charge of electricity in the air that causes pointed objects to glow when thunderstorms are in the area.

14. The scientific method requires observation. Because past weather conditions cannot be repeated or observed, we can only make educated guesses, which does not fulfill the scientific method.

15. A scientist must make an assumption about a past climate, create a model based on that assumption, and test how well data observed in the present fits the model.

16. Volcanic ash blocked sunlight from entering the atmosphere, causing cooler temperatures. The volcanoes also warmed the water, creating more evaporation and greater amounts of snowfall in northern regions.

17. 1 Thessalonians 5:21: "Examine everything carefully; hold fast to that which is good" (NASB).

18. Answers will vary, but possibilities include the heat island effect, changing instrument shelter location, placing shelters near heat sources, and changing instruments.

19. Hurricane landfalls, droughts, floods, tornadoes or thunderstorms, East Coast winter storms, heat waves, and cold spells.

20. Genesis 2:15: "The LORD God took the man and put him into the garden of Eden to cultivate it and keep it" (NASB).

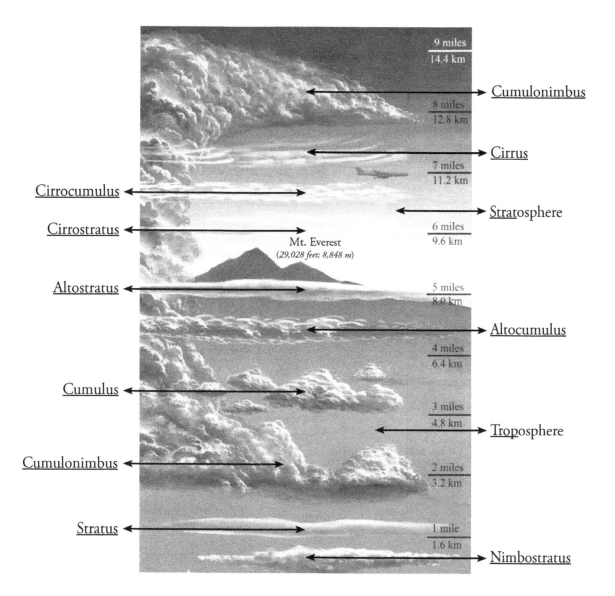

Cumulonimbus
Cirrus
Cirrocumulus
Stratosphere
Cirrostratus
Mt. Everest
(29,028 feet; 8,848 m)
Altostratus
Altocumulus
Cumulus
Troposphere
Cumulonimbus
Stratus
Nimbostratus

9 miles / 14.4 km
8 miles / 12.8 km
7 miles / 11.2 km
6 miles / 9.6 km
5 miles / 8.0 km
4 miles / 6.4 km
3 miles / 4.8 km
2 miles / 3.2 km
1 mile / 1.6 km

The New Astronomy Book ⚯ Unit Quiz Answer Key

Quiz One, Introduction – Chapter 3

1. astronomy – the study of heavenly bodies, things outside of the earth, including the sun, moon, and stars

2. spectroscopy – the study of spectra

3. comet – small bodies in space that contain frozen dust, gases, and even rocks, which have orbits of their own; made up of a nucleus, often with a trail of particles and dust that follow it

4. axis – an imaginary line, vertical and horizontal, around which a planet or other body rotates

5. retrograde motion – when a planet appears to move east to west with respect to the stars, opposite from its normal motion

6. maria – Latin for "seas"; refers to the darker areas of the moon's surface, at lower elevations; so called because astronomers once thought they might be bodies of water

7. highlands – areas of the craters on the surface of the moon that appear lighter and are at a high elevation

8. lunar – referring to features and aspects related to and upon the moon

9. satellites – another word to describe moons that orbit planets

10. minor planets – another name preferred by astronomers and given to bodies orbiting the sun that are

also called asteroids, smaller than planets

1. They appear to move clockwise around the south celestial pole; no, it does not.
2. A NEO is a near earth object.
3. The moon
4. Mercury, Venus. Mars, Jupiter, and Saturn
5. Rotation
6. When an object in space rotates on its axis at the same rate that it revolves around another body, like a planet; the moon
7. Galileo
8. It doesn't. It reflects light it gets from the sun.
9. The earth's gravity pulls on the moon, causing the moon to orbit the earth. But the moon's gravity also pulls on the earth and even alters its shape a little. The moon's gravity also impacts the ocean tides.
10. They knew the earth's shadow caused lunar eclipses, and that it was always circular. They also realized that only a sphere casts a circular shadow.
11. The corona Is the outermost layer of the sun; prominences are loops of gas that follow the sun's magnetic fields.
12. Pluto, Eris, Ceres, Haumea, Makemake
13. Mercury, Mars, Earth, Venus, Jupiter
14. Saturn, Uranus, Neptune, Pluto
15. Wearing out, collision with a planet, or being ejected from the solar system

Quiz Two, Chapters 4–7

1. Astronomical unit – a unit of distance, AU, which is based on the average distance between the earth and sun
2. Density – measure of how closely packed matter is, expressed in grams/cubic centimeter
3. Rotation periods – The period of time that it takes an object to spin on its axis once
4. Galilean satellites – refers to the four largest moons of Jupiter, named for the person who discovered them
5. parallax – the visual shift slightly back and forth each year of the stars, based on where we are viewing it from either side of earth's orbit
6. fission – a process where a large atom, such as uranium, fissions or breaks into smaller atoms, releasing a lot of energy
7. fusion – smaller atoms combining into larger atoms, creating energy
8. wavelengths – refers to how long a light wave is
9. electromagnetic radiation – the name for invisible wavelengths of light
10. infrared radiation – also known as heat radiation, refers to the radiation that has wavelengths longer than what is visible to the human eye
1. Because distances in space like miles and kilometers don't work well across vast areas; they would be enormous and impractical to use.
2. Venus and Uranus
3. terrestrial and jovian
4. terrestrial
5. Jovian

6. The gravitational pull of moons, like those of Saturn, tear rings apart — making their chance of surviving millions or billions of years after being formed very unlikely.

7. It traps the infrared radiation rather than letting it radiate away.

8. Mars

9. No; they think it has stayed closed to current temperatures.

10. Photosphere

11. They are regions of magnetic fields, and the sunspots in the pair have opposite polarity.

12. Refractors and reflectors

13. Different elements form different spectral lines so they can be used to tell the composition of the body.

14. The phases of Venus, craters on the moon, and four moons orbiting Jupiter

15. Eratosthenes; by measuring how high the sun was in the sky on the same date from two locations running along the north and south line. Then with the distance between the locations, the different in the height of the sun in the two locations, he calculated the circumference of the earth.

Quiz Three, Chapters 8-11

1. light years – the distance that light travels in a year; used to express distances between stars and other space objects

2. absolute magnitude – how bright a star would be if it were at the standard distance of 10 parsecs

3. parsec – a unit of measure of distance that astronomers use that is equal to 3.26 light years

4. white dwarf stars – stars that are a lot smaller than the sun, about the size of the earth, but they are so hot, they are white

5. extrasolar – planets that orbit a star beyond our solar system

6. transit – when a smaller body passes in front of a larger body; a type of eclipse where a much smaller body blocks out only a small portion of a bigger body's light

7. habitable zone – the theoretical range of distance between a star and an orbiting planet that might have liquid water on the surface

8. star clusters – groups of stars spaced much closer together than surrounding stars and held together by their gravity

9. interstellar medium – thin gas and other small materials that exist between stars

10. Local Group – the group of galaxies that includes the Milky Way and another spiral galaxy, M31

1. How bright a star appears to be to us

2. Its temperature

3. Hydrogen

4. Magnitude

5. Look for motion in the star as the planet orbits it

6. By decreasing the amount of the brightness we see

7. Less

8. Open and globular

9. Globular

10. Open

11. Yes

12. About 100,000 light years

13. Barred spiral galaxy

14. They were solar systems in development.

15. Virgo Cluster

Quiz Four, Chapters 12-15

1. dimension – a fundamental quantity of space or time, usually associated with a direction and measured in meters or seconds

2. *Ex Nihilo* – Latin word that means "out of nothing"; can refer to the creation, when God created everything from nothing

3. general relativity – the current theory of gravity, and how it relates space and time to matter and energy

4. Hubble relation – concept that redshifts and the distances of galaxies are related, such as the greater the redshift of a galaxy, the greater the distance of the galaxy

5. blueshifted – when the Doppler shift moves light to shorter wavelengths (toward blue), indicating objects are moving toward us

6. standard candle – an object for which we think we know the actual brightness, or absolute magnitude, that can be used to determine distances

7. quasars – small, high redshift objects that likely are very far away and very bright

8. synchrotron radiation – a special form of radiation created by very strong magnetic fields interacting with fast-moving charged particles

9. cosmology – the study of the structure of the universe

10. filaments – lines and flat or curved surfaces where galaxies seem to cluster

1. How we can see any galaxies if they are millions or billions of light years away if the universe is only 6,000 years old.

2. Going long distances in a short period of time, manning spacecraft for long voyages taking into account the average human life span, vehicle design that provides food, shelter, care, water, and scientific study equipment

3. Edwin Hubble in 1929

4. How fast or slow an object or source of light is moving toward or away from us

5. No

6. They were thought to be supermassive black holes.

7. 1950s

8. It is not as bright as those being fed large amounts of matter.

9. Yes, they are.

10. A physicist or astronomer who specializes in the study of cosmology or structure of the universe

11. The horizon problem and the flatness problem

12. Scientists changed the big-bang model to fit the data.

13. Project Moon-Blink

14. A larger expansion rate, inclusion of inflation and dark matter, dark energy, and the string theory

15. Most of what were thought to be nebulae of the time were redshifted.

The New Astronomy Book 🔭 Test Answer Key

1. parsec – a unit of measure of distance that astronomers use that is equal to 3.26 light years

2. Density – measure of how closely packed matter is, expressed in grams/cubic centimeter

3. minor planets – another name preferred by astronomers and given to bodies orbiting the sun that are also called asteroids, smaller than planets

4. white dwarf stars – stars that are a lot smaller than the sun, about the size of the earth, but they are so hot, they are white

5. infrared radiation – also known as heat radiation, refers to the radiation that has wavelengths longer than what is visible to the human eye

6. absolute magnitude – how bright a star would be if it were at the standard distance of 10 parsecs

7. blueshifted – when the Doppler shift moves light to shorter wavelengths, indicating objects are moving toward us

8. electromagnetic radiation – the name for invisible wavelengths of light

9. Astronomical unit – a unit of distance, which is based on the average distance between the earth and sun

10. retrograde motion – when a planet appears to move east to west with respect to the stars, opposite from its normal motion

11. axis – an imaginary line, vertical and horizontal, around which a planet or other body rotates

12. wavelengths – refers to how long a light wave is

13. maria – Latin for "seas"; refers to the darker areas of the moon's surface, at lower elevations

14. parallax – the visual shift slightly back and forth each year of the stars, based on where we are viewing it from either side of earth's orbit

15. highlands – areas of the craters on the surface of the moon that appear lighter and are at a high elevation

16. fusion – smaller atoms combining into larger atoms, creating energy

17. light years – the distance that light travels in a year; used to express distances between stars and other space objects

18. transit – when a smaller body passes in front of a larger body; a type of eclipse where a much smaller body blocks out only a small portion of a bigger body's light

19. star clusters – groups of stars spaced much closer together than surrounding stars and held together by their gravity

20. standard candle – an object for which we think we know the actual brightness, or absolute magnitude, that can be used to determine distances

1. They appear to move clockwise around the south celestial pole; no, it does not.

2. Mercury, Venus. Mars, Jupiter, and Saturn

3. When an object in space rotates on its axis at the same rate that it revolves around another body, like a planet; the moon

4. The earth's gravity pulls on the moon, causing the moon to orbit the earth. But the moon's gravity also pulls on the earth and even alters its shape a little. The moon's gravity also impacts the ocean tides.

5. The corona Is the outermost layer of the sun; prominences are loops of gas that follow the sun's magnetic fields.

6. Pluto, Eris, Ceres, Haumea, Makemake

7. Saturn, Uranus, Neptune, Pluto

8. Wearing out, collision with a planet, or being ejected from the solar system

9. Because distances in space like miles and kilometers don't work well across vast areas; they would be enormous and impractical to use.

10. Venus and Uranus

11. terrestrial and jovian

12. The gravitational pull of moons, like those of Saturn, tear rings apart — making their chance of surviving millions or billions of years after being formed very unlikely.

13. It traps the infrared radiation rather than letting it radiate away.

14. Mars

15. No; they think it has stayed closed to current temperatures.

16. They are regions of magnetic fields, and the sunspots in the pair have opposite polarity.

17. Different elements form different spectral lines so they can be used to tell the composition of the body.

18. How bright a star appears to be to us

19. Its temperature

20. Hydrogen

21. Magnitude

22. By decreasing the amount of the brightness we see

23. Less

24. Open

25. Yes

26. Barred spiral galaxy

27. How we can see any galaxies if they are millions or billions of light years away if the universe is only 6,000 years old.

28. How fast or slow an object or source of light is moving toward or away from us

29. It is not as bright as those being fed large amounts of matter.

30. The horizon problem and the flatness problem

Glossary

arid — a dry climate lacking moisture.

atmosphere — the body of gasses surrounding the earth.

avalanche — a rapid flow of snow down a sloping surface

axis — an imaginary straight line through the center of the earth on which it rotates.

ball lightning — a glowing ball of red, orange, or yellow light found during a thunderstorm.

barometer — a weather instrument used to measure the pressure of the atmosphere.

blizzard — a very heavy snowstorm with violent winds.

bogs — soft, waterlogged ground such as a marsh.

carbon dioxide — a colorless, odorless gas formed during respiration, combustion, and organic decomposition.

chinook winds — foehn winds that are mild, gusty, west winds found along the east slopes of the Rocky Mountains.

cirrus clouds — a high altitude cloud made of ice crystals that appear thin, white, and feathery.

climate — the weather conditions that are particular to a certain area, such as wind, precipitation, and temperature.

cold front — a boundary of cold air, usually moving from the north or west, which is displacing the warm air.

condensation — the act of water vapor changing from a gas to a liquid.

convection clouds — clouds that occur in a rising updraft, usually when the sun's radiation warms the earth. This causes the water vapor to condense.

Coriolis force — the movement of atmospheric air caused by the rotating earth.

creation — the formation of everything by God

cumulus clouds — low clouds that are thick, white, and puffy with flat bottoms and rounded tops.

dew point — the temperature at which air becomes saturated and dew forms.

Doppler radar — a special type of radar used to track severe weather by detecting wind speed and direction.

downdraft — a downward current of air.

dust devil — a relatively long-lived whirlwind on the ground formed on a clear, hot day

electricity — a moving electric charge, such as in a thunderstorm.

electrons — a subatomic particle with a negative electrical charge.

El Nino — a warm current from the west that replaces the cool ocean current along Peru and Ecuador.

environment — the circumstances or conditions around us.

environmentalist — someone concerned with the environment

equator — an imaginary line dividing the northern and southern hemispheres of the earth.

evaporation — to change into a vapor such as the evaporation of water by the warming of the sun.

flash flood — a flood caused by a thunderstorm that deposits an unusual amount of rain on a particular area.

foehn winds — a dry, warm downslope wind on the lee side of a mountain range

fog — clouds that form on the surface of the ground.

fossil fuels — coal and oil derived from the remains of plant and animal organisms.

frostbite — localized damage to skin and tissues due to freezing

funnel cloud — a rotating whirlwind below a cloud that has not yet touched the ground

greenhouse warming — the phenomena of a steady, gradual rise of temperatures due to the increase of carbon dioxide in the atmosphere. This could result in natural catastrophes such as droughts, flooding, and a meltdown of the ice sheets.

hailstones — precipitation in the form of ice and hard snow pellets.

heat-island effect — the warming effect in cities, compared to rural areas, caused by human activities

humid — a weather condition containing a large amount of moisture or water vapor.

hurricane — the strongest storm found in the tropics, with heavy rain and winds of 75 mph or greater.

hypothermia — a condition in which the body temperature drops below that required for biological functions

ice age — a period of time marked by extensive glaciers on the face of the earth.

ice cap — an extensive covering of ice and snow.

ice jam — the buildup of water caused by a blockage of ice

ice storm — a storm caused by rain falling into a lower atmosphere that is below freezing.

Intertropical Convergence Zone — area near the equator where winds from different directions merge or mix.

inversion — a condition in which the atmospheric temperature increases upward

lake-effect snowstorm — a snowstorm produced when cool air moves over long expanses of warm water, evaporating much moisture, and precipitating on downwind shores

La Niña — an ocean/atmospheric phenomenon that is the opposite of El Niño

latitudes — the distance north or south of the equator measured with imaginary lines on a map or globe.

Little Ice Age — a term used to refer to the global cool period between about A.D. 1300 to 1880

low-pressure system — Warm, moist air that usually brings storms with strong winds. The air spirals counter-clockwise around a low center in the Northern Hemisphere and clockwise in the Southern Hemisphere. Because the air is spiraling toward the center of the low, it is forced upward, forming clouds and precipitation.

Medieval Warm Period — a term used to refer to the warm period in the North Atlantic region, and possibly worldwide, from about A.D. 900 to 1200

meteorologist — a person that interprets scientific data and forecasts the weather for a specific area.

microburst — a small-scale intense downdraft from a thunderstorm

model — a conceptual representation of a phenomenon

monsoon — a wind system that causes periods of wet and dry weather in India and southern Asia.

nitrogen — a naturally occurring element that is responsible for around four-fifths of the earth's atmosphere.

Northeaster — a storm that moves northeast along the east coast.

oxygen — a colorless, odorless gas that is 21 percent of our atmosphere. It is essential for plant and animal respiration.

ozone — a gas in the earth's upper atmosphere that is responsible for screening most of the sun's harmful ultraviolet radiation.

permafrost — permanently frozen subsoil found around polar regions.

plankton — tiny plant and animal organisms found in the oceans.

pollution —harmful or unsafe waste products.

precipitation — falling moisture in the form of rain, sleet, snow, hail, or drizzle.

rain forest — a woodland with a high annual rainfall and very tall trees and that is often found in tropical regions

rain gauge — a weather instrument used to measure the amount of rainfall over a particular period of time.

relative humidity — the amount of water vapor in the air compared to the amount of water vapor the air can contain at the point of saturation.

Santa Ana winds — a foehn wind that blows westward from the mountains of southern California to the coast when a high pressure area settles over Nevada, Utah, and Idaho.

sedimentary rock — type of rock formed by the deposition of sediments at the earth's surface in water, and later cemented

sleet — precipitation that consists of frozen raindrops.

static electricity — a build-up of electrical charge on an insulated body.

St. Elmo's fire — a condition caused by a high charge of electricity in the air that causes pointed objects to glow slightly.

storm surge — a coastal flood caused by rising ocean water during a storm, in particular, storms in the tropics, like hurricanes

stratus clouds — low altitude gray clouds with a flat base.

subarctic — a region just south of the Arctic Circle.

supercell — a severe, well-organized thunderstorm with warm moist air spiraling upwards.

supercooled drops — drops of water that remain liquid below freezing

thermometer — an instrument used to indicate the temperature.

Theory of Evolution — a theory that claims that life came from non-life, new species arise through natural selection over millions of years.

thunderstorm — a condition of weather that produces thunder, lightning, and rain.

tide — a raising and lowering of the water in the oceans and seas caused by the gravitational pull of the moon. The sun causes some, but to a lesser degree.

tornado — a funnel-shaped column of air rotating up to 300 mph, touching the ground.

tornado alley — the term used in the United States where tornadoes are more frequent, centered in northern Texas, Oklahoma, Kansas, and Nebraska

trade winds — the prevailing pattern of easterly winds in the tropics

tropical — a warm climate located near the equator, usually having lots of precipitation.

tropical depression — rainstorms with winds of 38 mph or less.

tropical storm — a storm of heavy rain and winds between 39 and 74 mph.

tundra — a region usually located at high altitude. The ground is permanently frozen.

typhoon — another name for a hurricane.

ultraviolet light — the range of wavelengths just beyond violet in the visible spectrum. Invisible to humans, yet capable of causing skin cancer.

uniformitarianism — the assumption that the same natural laws we observe today have always operated in the past, and is remembered by the saying "the present is the key to the past"

updraft — an upward current of air.

warm front — a boundary of warm air which pushes out cold air in the atmosphere.

waterspout — a tornado over a body of water

water vapor — invisible water distributed throughout the atmosphere.

weather balloons — balloons used to carry weather instruments into the atmosphere to gather data.

weather vane — an instrument used to indicate wind direction.

wind chill factor — the temperature of windless air that would have the same cooling effect on exposed skin as a combination of wind speed and air temperature.

absolute magnitude — how bright a star would be if it were at the standard distance of 10 parsecs.

ASC solution — stands for the anisotropic synchrony convention solution by Dr. Lisle to the problem of light-travel-time.

asteroids — considered to be minor planets, especially if located in the inner region of our solar system.

astronomical — a part of or related to aspects within astronomy; also used to describe extremely large distances or amounts.

astronomical unit — a unit of distance, AU, which is based on the average distance between the earth and sun.

astronomy — the study of heavenly bodies, things outside of the earth, including the sun, moon, and stars.

astrophysics — the application of modern physics to the study of astronomy.

axis — an imaginary line, vertical and horizontal, around which a planet or other body rotates.

big-bang theory — a popular but unproven idea that the universe began about 13.8 billion years ago in a very hot, dense state and has been expanding and cooling since.

binary systems — binary, meaning, two; a star system in which two stars are orbiting one another.

black hole — a star that is so massive and small that its gravity is very strong, so strong that light cannot escape.

blueshifted — When the Doppler shift moves light to shorter wavelengths (toward blue), indicating objects are moving towards us.

bombardment — when an object in space, like the moon, has been struck by many other objects (comets, meteors, etc.) in space.

celestial — a reference term related to the universe and objects within it.

cepheid variables — are pulsating giant and supergiant stars that follow a period-luminosity relationship, making them useful for measuring distances.

circumpolar — means "around the pole," referring to stars that from a given location neither rise nor set but appear to circle the pole.

coma — the gas and dust that expand to form a large cloud when a comet passes close to the sun.

comet — small bodies in space that contain frozen dust, gases and even rock which have orbits of their own; made up of a nucleus often with a trail of particles and dust that follow it.

constellations — a group of stars that seem to form a pattern or shape.

continuous-creation theory — alternate name for the steady-state theory, an idea that matter popped into existence to keep a constant density in the universe.

cosmic microwave background — radiation of a particular type in the microwave part of the spectrum that is coming from all directions.

cosmology — the study of the structure of the universe.

cryovolcanism — volcanic activity that takes place on very cold objects in space.

dark nebula — a dense dust cloud in space that blocks the light of stars and so appears dark.

dasha solution — a solution to the light-travel-time problem that suggests that God miraculously and quickly brought the light of distant stars to the earth on day 4 of the creation week.

density — a measure of how closely packed matter is, expressed in grams/cubic centimeter.

dimension — a fundamental quantity of space or time, usually associated with a direction and measured in meters or seconds.

Doppler Effect — the shift in spectral lines from their normal wavelengths that help to show whether an object in space is moving toward or away from us.

dwarf galaxies — very small galaxies, many of which orbit much larger galaxies as a satellite.

electromagnetic radiation — the name for invisible wavelengths of light.

ellipse — shaped like a flattened circle, referring to the shape of some galaxies that do not have disks or spiral arms.

epicycle — a smaller circle that moved on a larger circle that Ptolemy used to explain retrograde motion of planets.

ex nihilo — Latin word that means "out of nothing"; can refer to the creation, when God created everything from nothing.

exoplanet — another name for extrasolar planets, or planets that are believed to be orbiting a star outside of our solar system.

extra-galactic astronomy — the study of other galaxies other than the Milky Way.

extrasolar — planets that orbit a star beyond our solar system.

filaments —lines and flat or curved surfaces where galaxies seem to cluster.

fission — a process where a large atom, such as uranium, fissions or breaks into smaller atoms, releasing a lot of energy.

flatness problem — a problem in the big-bang model that refers to the certain type of geometry that the universe has.

fusion — smaller atoms combining into larger atoms, creating energy.

Galilean satellites — refers to the four largest moons of Jupiter, named for Galileo who discovered them.

general relativity — the current theory of gravity, and how it relates space and time to matter and energy.

geocentric theory — means "earth-centered," or the idea that the earth remained motionless while the sun orbited it.

globular clusters — a type of star cluster that usually is spherical, with many, many stars — 50,000 up to a million.

habitable zone — the theoretical range of distance between a star and an orbiting planet that might have liquid water on the surface.

heliocentric theory — means "sun-centered," or the theory that planets of our solar system orbit around the sun.

highlands — areas of the craters on the surface of the moon that appear lighter and are at a high elevation.

horizon problem — a problem with the big-bang model in that the cosmic microwave background has the same temperature in every direction.

Hubble relation — a concept that redshifts and the distances of galaxies are related, such as the greater the redshift of a galaxy, the greater the distance of the galaxy.

impact basins — refers to large, round, dark features on the surface of the moon, created by impacts of large space objects of some kind.

infrared radiation — also known as heat radiation, refers to the radiation that has wavelengths longer than what is visible to the human eye.

interstellar medium — thin gas and other small materials that exist between stars.

irregular galaxies — galaxies that cannot be placed into the common categories of galaxies like spiral or elliptical.

Jovian — means Jupiter-like, refers to planets with certain characteristics — farther from the sun, more mass, and much larger.

laws of planetary motion — three laws that describe the motions of the planets discovered by Johannes Kepler.

light years — the distance that light travels in a year; used to express distances between stars and other space objects.

Local Group — the group of galaxies that includes the Milky Way and another spiral galaxy, M31.

lunar — referring to features and aspects related to and upon the moon.

magnetic polarity — refers to having a magnetic north or south, as sunspots have.

main sequence star — a grouping of stars rated as normal like the sun, can be as small as 1/10th the size of the sun, up to 10 time larger than the sun.

maria — Latin for "seas," refers to the darker areas of the moon's surface, at lower elevations, so called, because astronomers once thought that they might be bodies of water.

mass — the measurement of how much matter a body (planet, star, etc.) may have.

microwaves — a part of the electromagnetic spectrum beyond infrared but before the radio part of the spectrum.

minor planets — another name preferred by astronomers and given to bodies orbiting the sun that are also called asteroids, smaller than planets.

molecular clouds — the largest and densest regions of the interstellar medium, containing molecules as well as atoms and dust particles.

nebulae — what are thought to be clouds of glowing gas in space.

neutron star — a very dense star that is about the size of the earth but has several times the mass of the sun.

nucleus — the center portion of a comet, usually made of ice and dust.

open clusters — one of two types of star clusters, usually with fewer stars and an irregular appearance.

orbital period — the length of time required for an orbiting body to complete one orbit.

pagan — refers to ancient cultures that were not based on Christianity, Islam, or Judaism.

paradox — an idea that seems contradictory, so requires an explanation.

parallax — the visual shift slightly back and forth each year of the stars, based on where we are viewing it from either side of earth's orbit.

parsec — a unit of measure of distance that astronomers use that is equal to 3.26 light years.

penumbra — the partial shadow of the earth on the moon.

photosphere — refers to what appears to be the surface area of the sun.

physics — the study of motion, energy, matter, and force.

planet — a large celestial body that is orbiting around a star.

potential energy — the energy that an object has because of its location.

pulsar — another name for neutron stars that appear to pulse because of flashes of radiation seen.

quasars — small, high redshift objects that likely are very far away and very bright.

quasi-stellar objects — another name for quasars.

radio stars — the name used for quasars before their nature was better known, so called, because they looked like stars but gave off radio emissions.

radio waves — the longest wavelength electromagnetic radiation.

redshifted — When the Doppler shift moves light to longer wavelengths (toward red), indicating objects are moving away from us.

reflection nebula — a nebula where the light of hot, bright stars near the nebula scatters, or reflects off the dust and appears blue in images.

retrograde motion — when a planet appears to move east to west with respect to the stars, opposite its normal motion.

revolution period — the length of time it takes a body in space to complete an orbit around another object, for example the earth around the sun.

rotation period — The period of time that it takes an object to spin on its axis once.

satellites — the length of time that it takes an object, such as the earth, to spin once on its axis.

silicate — a type of rock.

spectral lines — unique patterns of absorption or emission lines that help scientists determine the composition of astronomical bodies in space.

spectrograph — a device or program that divides light up into its wavelengths, or colors.

spectroscopy — the study of spectra.

spiral galaxies — two types of galaxies that have the characteristic of spiral arms.

spiral nebulae — what astronomers used to call what are now known as spiral galaxies.

spiral tracers — objects that lie along spiral arms of galaxies, and so can be used to trace the locations of spiral arms.

standard candle — an object for which we think we know the actual brightness, or absolute magnitude, that can be used to determine distances.

star clusters — groups of stars spaced much closer together than surrounding stars and held together by their gravity.

steady-state theory — see continuous-creation theory definition.

sun – a star at the center of a solar system providing light and heat, around which planets orbit.

sunspots — areas that appear to be dark spots on the photosphere of the sun.

supernovae — the very bright explosions of certain stars that astronomers think are the end of such stars.

synchrotron radiation — a special form of radiation created by very strong magnetic fields interacting with fast-moving charged particles.

synchrotron spectra — a unique kind of spectrum produced by sources of synchrotron radiation.

terrestrial — means "earth-like"; also refers to the four closest planets to our sun.

transit — when a smaller body passes in front of a larger body, a type of eclipse with a much smaller body blocks out only a small portion of the bigger body's light.

unmanned — refers to missions or spacecraft that do not have humans onboard to operate them in space.

umbra — a shadow of an astronomical body in which the light source has been completely blocked out, such as occurs during a total eclipse.

voids — areas between clusters of galaxies that appear to be open or empty.

wavelengths — refers to how long a light wave is.

white dwarf stars — stars that are a lot smaller than the sun, about the size of earth, but they are so hot, they are white.

white hole theory — (see page 73) The theory based on the idea that matter and energy fly outward from a white hole. As matter and energy exit a white hole, the white hole would get smaller and eventually go away. However, time would pass at very different rates inside and outside the white hole. If the earth is inside the white hole and stars are outside, millions of years might pass for the stars and the space outside, but only a few days might have passed on earth.